Mathematics

YEAR 6

Serena Alexander

GALORE PARK

AN HACHETTE UK COMPANY

The publishers would like to thank the following for permission to reproduce copyright material:
Photo credits p132 © Piotr Marcinski – Fotolia.com **p134(t)** © Ros Drinkwater / Alamy
p134(b) © Alexander Kaludov – Fotolia.com **p192** © Benoît Crépin – Fotolia.com

Acknowledgements p138 © Crown copyright 2014. Ordnance Survey Licence number 150001477
p211 © British Crown Copyright and/or database rights. Reproduced by permission of the Controller
of Her Majesty's Stationery Office and the UK Hydrographic Office (www.ukho.gov.uk).

Although every effort has been made to ensure that website addresses are correct at time of going to
press, Galore Park cannot be held responsible for the content of any website mentioned in this book.
It is sometimes possible to find a relocated web page by typing in the address of the home page for a
website in the URL window of your browser.

Hachette UK's policy is to use papers that are natural, renewable and recyclable products and made
from wood grown in sustainable forests. The logging and manufacturing processes are expected to
conform to the environmental regulations of the country of origin.

Orders: **Teachers** please contact Bookpoint Ltd, 130 Park Drive, Milton Park, Abingdon, Oxon OX14
4SE. Telephone: (44) 01235 400555. Email primary@bookpoint.co.uk. Lines are open from 9 a.m. to
5 p.m., Monday to Saturday, with a 24-hour message answering service.

Parents, Tutors please call: 020 3122 6405 (Monday to Friday, 9:30 a.m. to 4.30 p.m.).Email:
parentenquiries@galorepark.co.uk

Visit our website at www.galorepark.co.uk for details of other revision guides for Common Entrance,
examination papers and Galore Park publications.

ISBN: 978 1 471829 36 9

© Serena Alexander 2014
First published in 2014 by Galore Park Publishing Limited
An Hachette UK Company
Carmelite House
50 Victoria Embankment
London EC4Y 0DZ
www.galorepark.co.uk

Impression number 10 9 8 7 6 5
2020   2019   2018

Typeset in India
Printed in Dubai
Illustrations by Aptara, Inc.

A catalogue record for this title is available from the British Library.

About the author

Serena Alexander has taught mathematics since 1987, originally in both maintained and independent senior schools. From 1999 she taught at St Paul's School for Boys, where she was Head of mathematics at their Preparatory School, Colet Court, before moving first to Newton Prep as Deputy Head and more recently to Devonshire House. She is an ISI inspector and helps to run regular mathematics conferences for prep school teachers. She has a passion for maths and expects her pupils to feel the same way. After a lesson or two with her, they normally do!

Contents

Introduction

This book is for pupils in Year 6. The author aims to provide a sound and varied foundation on which pupils can build in the future. There is plenty of material to support this but, at the same time, there are possibilities for the more able to be extended.

The author does not wish to dictate to either pupil or teacher. A combination of approaches, the more modern 'mental' and the more historical 'traditional', are both explored, so that the appropriate method for the individual can be adopted.

There is no prescribed teaching order. Topics may well be taught more than once during the year. The author is convinced that it is the teacher who knows what is best for each individual pupil – and when each topic should be introduced.

⇨ Notes on features

Words printed in blue and bold are keywords. All keywords are defined in the Glossary at the end of the book.

> Useful rules and reminders looking like this are scattered throughout the book.

Worked examples are given throughout to aid understanding of each part of a topic.

Activity

Mathematics is so often a question of patterns. Many chapters end with a freestanding activity, either numerical or spatial, to cover this aspect of the subject.

For some questions and activities, pupils are asked to copy diagrams from the book. They may find tracing paper helpful when doing this. Such activities are also supported by separate worksheets. These worksheets may be photocopied from the section at the back of the answers (available separately).

① Investigations with numbers

When you are asked a question in mathematics, it may not be obvious how to answer it simply by calculation. You will need to study the problem and try to identify patterns, then use what you have found out to reach a solution. Many investigations will lead to a **general solution** that can be expressed in words or letters.

⇨ Shaking hands

Just before a big event in the 2014 Winter Olympics, 40 competitors from around the world were gathered at the top of the mountain.

One turned to another and said: 'I think I will go and shake hands with all of the competitors and wish them good luck. I shall encourage them all to shake everyone else's hand!'

Her friend said: 'Are you serious? Do you know how many handshakes there would be, altogether?'

If all 40 of the competitors were to shake hands with everyone else, how many handshakes would there be in total?

To solve this puzzle, look at some simpler situations and see if you can spot a pattern.

One person alone cannot shake hands with anyone else.

So the total number of handshakes is zero!

If there are two people, they can shake hands with each other.

The total number of handshakes is 1

Now suppose there are three people, Adam, Ben and Carla.

A B   A C   C B

Adam and Ben shake hands, Ben and Carla shake hands and Adam and Carla shake hands.

Is that all? How many handshakes will there be altogether?

Exercise 1.1

1 How many handshakes will there be, if there are four people? Work out how many handshakes there will be for four people, five people and six people. Use diagrams like these and count how many ways there are of joining two dots.

4 people

5 people

6 people

2 Copy and complete this table.

Number of people	Number of handshakes
2	1
3	3
4	
5	
6	

3 Look at the pattern of numbers. Continue the pattern and work out the number of handshakes there will be for 7 people and for 8 people.

Number of people	Number of handshakes
7	
8	

4 Copy these diagrams to work out the number of handshakes for 7 people and for 8 people, to check your answer to question 4

7 people

8 people

5 Continue the pattern of numbers to find the numbers of handshakes there will be for 9 people and for 10 people.

Number of people	Number of handshakes
9	
10	

⇨ Triangular numbers

The numbers 1, 3, 6, 10, 15, 21... are called **triangular numbers**.

The pattern below shows you why.

•   •   •   •
    •• •• ••
        ••• •••
             ••••

1 = 1   1 + 2 = 3   1 + 2 + 3 = 6   1 + 2 + 3 + 4 = 10

Each triangular number is the sum of a sequence of **consecutive** numbers, starting with one. **Consecutive numbers** are numbers that follow each other in order; for example, **5, 6 and 7 are consecutive numbers.**

Therefore the 10th triangular number is the sum of all the numbers from 1 to 10.

$$1 + 2 + 3 + 4 + 5 + 6 + 7 + 8 + 9 + 10 = 55$$

Triangular numbers occur frequently in mathematics.

You can work out triangular numbers by continuing to add consecutive numbers, but it would be easier if there were a formula.

One rule or formula could be:

nth triangular number $= 1 + 2 + ... + n$

This is not very useful, though. You need a more general formula.

1 Draw the first four triangular numbers on a squared grid, like this. Here are the first three to start you off.

Leave at least three squares between the numbers.

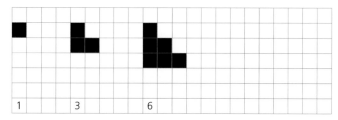

2 Now repeat each triangular number, upside down, next to itself, to make a **rectangle**. Do this for each number.

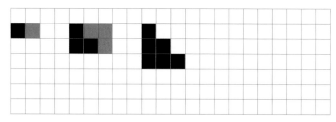

3 Copy and complete these calculations.

$1 + 1 = 2$ $3 + 3 = 6$ $6 + 6 = ...$ $10 + 10 = ...$

4 Draw the next three triangular numbers.

Draw each one again, upside down, next to the first, to make a rectangle.

Pattern 5 will look like this.

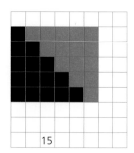

5 Look at the height and width of each rectangle you have made.

Copy and complete this table.

Pattern number	Triangular number	Height	Width	Area (2 × triangular number)
1	1	1	2	2
2	3	2	3	6
3	6	3		12
4	10	4		
5				
6				
7				
10				
n				

6 Explain, in words, how you could find the area of the rectangle made up of two identical triangular numbers, each of height n.

7 Write a rule or formula for the nth triangular number.

More triangular numbers

The next exercise is also about adding triangular numbers, but this time you will add **consecutive** triangular numbers. Consecutive numbers are numbers that follow each other, so 2 and 3 are consecutive numbers. Then 3 and 6 are **consecutive triangular numbers**.

1 Draw the first four triangular numbers on a squared grid. Here are the first two to start you off.

Leave at least three squares between the numbers.

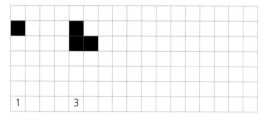

2 Now, on each one, draw the previous triangular number, upside down, on top of the first to make a square.

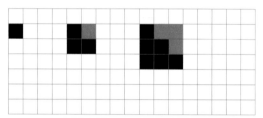

3 Copy and complete these calculations.

$1 + 0 = 1$ $3 + 1 = 4$ $6 + 3 = ...$ $10 + 6 = ...$

4 Draw the next three triangular numbers. For each one, add the previous triangular number, upside down, to make a square (pattern number 5 is drawn below).

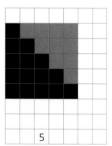

5 Copy this table. Look at each square and complete your table.

Pattern number	First triangular number	Second triangular number	Total area
1	1	0	1
2	3	1	4
3	6	3	9
4	10	6	
5			
6			
7			
10			
n			

Do you recognise the numbers in the last column? They are called square numbers because they can be represented by squares of dots. A square number is the result of multiplying a number by itself.

6 Copy and complete this sentence.

The sum of two consecutive triangular numbers is a ...

⇨ Patterns in numbers

When you look closely at square numbers and triangular numbers, you can identify many interesting patterns. This next exercise will show you another pattern.

Exercise 1.4

1 Copy this sequence of patterns into your book. Draw the next three patterns in the sequence.

2 Copy and complete this table.

Pattern	Black	Red	Total
1	1	0	1
2	1	3	4
3	6	3	9
4		10	
5			
6			
7			

(a) Look at the numbers in the totals column. What sort of numbers are they?

(b) Look at the numbers in the black and red columns. What sort of numbers are they?

3 Extend your table up to pattern number 10

4 Write down any patterns that you notice. Then write a rule for finding the number of red circles and the number of black circles in the nth pattern of the sequence.

2 Working with numbers

⇨ Place value

The system that you use for counting is based on the ten **digits** 1, 2, 3, 4, 5, 6, 7, 8, 9 and 0. For any digit, its **place** within the number tells you its **value**.

In 719, the 7 stands for 7 hundred (700) and the number is seven hundred and nineteen.

H	T	U
7	1	9

In 702 120 the 7 stands for 7 hundred thousand (700 000) and the number is seven hundred and two thousand, one hundred and twenty.

Thousands			HTU		
HTh	TTh		H	T	U
7	0	2	1	2	0

*The thick line shows the thousands are in a different block from HTU.

In 175 050 009 the 7 stands for 70 million and the number is one hundred and seventy-five million, fifty thousand and nine.

Millions			Thousands			HTU		
HM	TM		HTh	TTh		H	T	U
1	7	5	0	5	0	0	0	9

*The thick line shows the millions are in a different block from the thousands.

If a number has more than four digits, you should use a small space to separate the tens and hundreds of thousands from the hundreds, tens and units.

- 8750 – eight thousand, seven hundred and fifty (no space)

- 12 415 – twelve thousand, four hundred and fifteen (space)

1 Write each of these numbers in figures.

(a) Four hundred and fifty

(b) Fifty thousand, six hundred and twenty-five

(c) Six hundred and five thousand, seven hundred and twelve

(d) Five million, six hundred and fourteen thousand and ninety

(e) Sixty-five million, twenty thousand, nine hundred and one

(f) One hundred and twenty million, six hundred and one thousand and fifty

2 Write each of these numbers in words.

(a) 901

(b) 1345

(c) 25 609

(d) 345 718

(e) 2 345 581

(f) 25 304 603

(g) 107 450 345

(h) 340 212 070

(i) 45 605 000

(j) 740 670 407

3 Write down the value of the underlined digit in each number.

(a) 6<u>5</u>0

(b) 193<u>5</u>

(c) 4<u>5</u>078

(d) <u>5</u>12 706

(e) <u>5</u>1 312 608

(f) 4<u>5</u>607 040

(g) 312 4<u>5</u>2 609

(h) 10<u>0</u>070 600

(i) 3 40<u>5</u>216

(j) 41 <u>5</u>00 095

⇨ Below zero

When you count, you generally start with one, but there are numbers that are less than one. If you counted backwards from 5, you would say:

five, four, three, two, one, zero, minus one, minus two, minus three and so on.

You can see this on a **number line**.

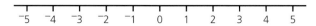

This number line counts in twenties.

Exercise 2.2

Draw these number lines.

1 From ⁻10 to 2, counting in ones.

2 From 0 to 100, counting in tens.

3 From ⁻50 to 50, counting in tens.

4 From ⁻100 to 40, counting in twenties.

5 From ⁻800 to 200, counting in hundreds.

6 From ⁻90 to 90, counting in twenties.

7 From ⁻100 to 300, counting in forties.

8 From ⁻1000 to 2200, counting in four hundreds.

9 From ⁻500 to 50, counting in fifties.

10 From ⁻1000 to 3000, counting in five hundreds.

⇨ Ordering and comparing

When you compare numbers, you can use the symbol > for **greater than** and < for **less than**.

2 300 000 < 3 200 000   This means: 'two million, three hundred thousand is less than three million two hundred thousand'.

15 070 000 > 13 405 000   This means: 'fifteen million and seventy thousand is greater than thirteen million, four hundred and five thousand'.

When you put numbers in order they can be in either **ascending** order or **descending** order.

- For **ascending** order, write the numbers from lowest (first) to highest (last) – think of ascending a mountain.

- For **descending** order, write the numbers from highest (first) to lowest (last) – think of making a descent.

When you are ordering numbers, follow these steps.

- Work out the value of the first digit in each number.

- If the first digits have the same value, then look at the second digits and work out their values.

- If the second digits also have the same value, then look at the third digits and work out their values.

- Continue until you have worked out the largest number.

Examples:

(i) Write these numbers in ascending order.

    1975   1957   1759   1795

(ii) Write these numbers in descending order.

    10 907 050   10 790 500   10 970 005   10 790 005

(i) 1759, 1795, 1957, 1975

(ii) 10 970 005, 10 907 050, 10 790 500, 10 790 005

Exercise 2.3

1 Copy each pair of numbers and write > or < between them. Then write the number sentence in words.

(a) 2014 ☐ 2041

(b) 12 405 ☐ 12 504

(c) 135 605 ☐ 135 506

(d) 1 305 407 ☐ 1 304 507

(e) 340 405 ☐ 351 504

(f) 5000 ☐ 4999

(g) 999 999 ☐ 1 000 000

(h) 13 599 999 ☐ 13 600 000

2 Write the numbers in each set in ascending order.

(a) 125   ⁻12   ⁻100

(b) 3003   3300   3033

(c) 99 909   9999   99 009

(d) 1 405 435   1 450 435   1 405 345

(e) 18 543 652   18 543 256   18 543 562   18 543 265

(f) 29 879 999   29 897 998   29 879 998   29 897 999

3 Write the numbers in each set in descending order.

(a) ⁻50   ⁻75   75

(b) 9099   9909   9009

(c) 501 345   503 145   503 345

(d) 5 550 000   5 555 000   5 550 005

(e) 10 900 900   10 900 009   10 900 909   10 909 009

(f) 450 050 000   450 500 050   45 050 005   450 500 005

4 Draw a time line from 100 B.C. to 150 A.D. Then put these dates on it.

54 B.C.	Julius Caesar launched an attack on Britain.
122 A.D.	Hadrian ordered a wall to be built across Northern Britain.
43 A.D.	Britain was conquered by the Romans.
100 B.C.	First coins were minted in Britain.

5 Pupils in our class have been measuring the number of steps they each take every day. These are the results from my group.

Name	Number of paces
Adam	9706
Ben	10 457
Charlie	10 047
Darren	10 754
Eva	9605

Rewrite the table in order of the number of paces.

6 This table shows the distances from the Sun of the five planets that are closest to it.

Planet	Distance from the Sun (km)
Earth	149 600 000
Jupiter	778 330 000
Mars	227 940 000
Mercury	57 910 000
Venus	108 200 000

(a) Rewrite the table in order of distance from the Sun.

(b) Which planet is closest to the Earth?

⇨ Rounding numbers

Looking at the table of planets, you could work out the order they should be in just by looking at the millions of kilometres.

Venus is 108 million kilometres from the Sun.

It looks as if Mercury is 57 million kilometres from the Sun, but is it?

Look at this number line.

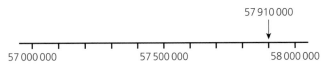

Mercury is 57 910 000 kilometres from the Sun. That is between 57 million and 58 million – but it is closer to 58 million. You can round the distance of Mercury from the Sun to 58 million kilometres.

When rounding any number, remember:

● If the next digit falls below the halfway mark (0, 1, 2, 3 or 4), round down.

● If the next digit falls at or above the halfway mark (5, 6, 7, 8 or 9), round up.

● When rounding a number do not use the equals sign but use ≈, which means 'is approximately equal to'.

Examples:

(i) Write each of these numbers correct to the nearest hundred.

   (a) 561   (b) 1325   (c) 453 750

(ii) Write each of these numbers correct to the nearest hundred thousand.

   (a) 720 000   (b) 750 000   (c) 780 000

(iii) Write each of these numbers correct to the nearest million.

   (a) 14 734 000   (b) 9 500 000   (c) 5 250 000

(i) (a) 561 ≈ 600   Above the halfway mark of 550

   (b) 1325 ≈ 1300   Below the halfway mark of 1350

   (c) 453 750 ≈ 453 800   At the halfway mark of 453 750

(ii) (a) 720 000 ≈ 700 000   Below the halfway mark of 750 000

   (b) 750 000 ≈ 800 000   At the halfway mark of 750 000

   (c) 780 000 ≈ 800 000   Above the halfway mark of 750 000

(iii)(a) 14 734 000 ≈ 15 000 000   Above the halfway mark of 14 500 000

   (b) 9 500 000 ≈ 10 000 000   At the halfway mark of 9 500 000

   (c) 5 250 000 ≈ 5 000 000   Below the halfway mark of 5 500 000

Exercise 2.4

1 Round each of these numbers to the nearest twenty.

   (a) 1458   (b) 23 455   (c) 305 123

2 Round each of these numbers to the nearest fifty.

   (a) 5360   (b) 14 325   (c) 345 620

3 Round each of these numbers to the nearest hundred.

   (a) 75 040   (b) 435 350   (c) 1 452 990

4 Round each of these numbers to the nearest ten thousand.

   (a) 2 456 000   (b) 2 445 000   (c) 2 453 000

5 Round each of these numbers to the nearest hundred thousand.

(a) 10 590 000   (b) 12 550 000   (c) 16 445 000

6 This table shows the distances from the Sun of the five planets that are closest to it.

Planet	Distance from the Sun (km)
Earth	149 600 000
Jupiter	778 330 000
Mars	227 940 000
Mercury	57 910 000
Venus	108 200 000

Rewrite each distance to the nearest million kilometres.

7 The *Gaia* spacecraft is in orbit 932 056 miles from the Earth. What is this to:

(a) the nearest ten thousand miles

(b) the nearest hundred thousand miles

(c) the nearest million miles

(d) the nearest fifty miles

(e) the nearest twenty miles?

⇨ Estimating

When you are talking about numbers, you do not always need to give exact values. Think about the size of a football crowd, or the distances between planets. A number that is given to the nearest thousand, hundred thousand or million is close enough.

Rounding numbers allows you to make an estimate of a calculation. Then you can work it out mentally.

Example:

How many millions of miles closer to the Sun is Earth (149 600 000 miles from Sun) than Mars (227 940 000 miles from Sun)?

The distance is 227 940 000 **minus** 149 600 000

Estimating these distances to the nearest million:

227 940 000 $\approx$ 228 million

149 600 000 $\approx$ 150 million

Then 228 million minus 150 million = 78 million

Earth is 78 million miles closer to the Sun than Mars.

Exercise 2.5

1 Estimate the answer to each calculation. Give each answer correct to the nearest thousand.

   (a) 24 712 + 54 356   **(c)** 17 543 + 89 501

   (b) 124 576 − 91 345   **(d)** 65 432 − 8795

2 Estimate the answer to each calculation. Give each answer correct to the nearest ten thousand.

   (a) 784 561 + 543 760   **(c)** 1 456 765 + 3 405 672

   (b) 732 125 − 464 654   **(d)** 1 435 658 − 612 345

3 Estimate the answer to each calculation. Give each answer correct to the nearest hundred thousand.

   (a) 1 345 230 + 435 900   **(c)** 1 974 657 + 2 435 890

   (b) 1 324 690 − 543 768   **(d)** 5 536 734 − 1 352 682

4 The table shows the figures, taken from the 2011 census, for the population and area of each of the five largest UK cities.

	City	2011 census Population
1	London	9 787 426
2	Manchester	2 553 379
3	Birmingham	2 440 986
4	Leeds	1 777 934
5	Liverpool	864 122

Estimate to the nearest hundred thousand, how many people live in:

(a) London and Manchester altogether

(b) Birmingham and Leeds altogether

(c) Manchester and Liverpool altogether

(d) London and Liverpool altogether.

5 This is a list of the tallest mountains in some European countries.

Country	Mountain	Height in m
Russia	Mount Elbrus	5642
France	Mont Blanc	4810
Austria	Grossglockner	3798
Andorra	Coma Pedrosa	2942
Greece	Mount Olympus	2919

(a) To the nearest one hundred metres, how much higher is:

   (i) Mount Elbrus than Grossglockner

   (ii) Mont Blanc than Mount Olympus

   (iii) Grossglockner than Coma Pedrosa?

(b) To the nearest ten metres, how much higher is:

   (i) Mount Elbrus than Mount Olympus

   (ii) Mont Blanc than Coma Pedrosa

   (iii) Grossglockner than Mount Olympus?

6 This is a list of the numbers of words in some well-known books.

Book	Author	Words
Charlotte's Web	E. B. White	14 458
The Hobbit	J. R. R. Tolkien	95 022
Harry Potter and the Philosopher's Stone	J. K. Rowling	76 944
The Lion, the Witch and the Wardrobe	C. S. Lewis	37 492
Matilda	Roald Dahl	40 309

(a) To the nearest thousand, how many more words are there in:

  (i) *The Hobbit* than *The Lion, the Witch and the Wardrobe*

  (ii) *Harry Potter and the Philosopher's Stone* than *Matilda*

  (iii) *Matilda* than *Charlotte's Web*?

(b) To the nearest hundred, how many words altogether are there in:

  (i) *Harry Potter and the Philosopher's Stone* and *The Lion, the Witch and the Wardrobe*

  (ii) *The Lion, the Witch and the Wardrobe* and *Charlotte's Web*

  (iii) *The Hobbit* and *Charlotte's Web*?

⇨ Mental arithmetic

Adding and subtracting

In the last exercise, you had to add and subtract **mentally** to find your answers. Look at these calculations.

$$144 - 61$$

$$98 + 26$$

Some people do this in their heads, in the same way they would if the calculation was written down, taking each column in turn, from the units up to the hundreds.

Other people imagine a number line and break the calculation into stages.

98 + 26 = 98 + 2 + 24 = 124

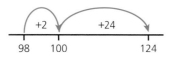

Another way is to break one number into its hundreds, tens and units.

144 − 61 = 144 − 60 − 1

= 84 − 1

= 83

It is always a good idea to stop and check that your answer looks sensible, before you move on to the next question.

There is no one correct way of calculating mentally. If it works for you then it is right! You do not need to write down all the stages. Just follow them through in your head.

For subtraction, you can check your answer by using the inverse.

144 − 61 = 83 then 61 + 83 = 144

Exercise 2.6

1 (a) Calculate the answers mentally.

   (i) 204 + 45   (iii) 126 + 144   (v) 675 + 95

   (ii) 350 + 155   (iv) 536 + 364   (vi) 867 + 133

(b) Write down an explanation for how you worked out your answer to 867 + 133

2 (a) Calculate the answers mentally.

   (i) 1035 + 145   (iii) 3209 + 801   (v) 1789 + 341

   (ii) 2034 + 176   (iv) 1534 + 172   (vi) 1932 + 158

(b) Write down an explanation for how to work out 1765 + 235

3 (a) Calculate the answers mentally.

   (i) 136 − 24   (iii) 735 − 64   (v) 538 − 62

   (ii) 348 − 49   (iv) 345 − 37   (vi) 712 − 95

(b) Write down an explanation for how to work out 735 − 196

4 (a) Calculate the answers mentally.

    (i) 1307 − 315   **(iii)** 1006 − 75   **(v)** 1305 − 947

    (ii) 1401 − 354   **(iv)** 4163 − 345   **(vi)** 1623 − 548

  (b) Write down an explanation for how to work out 1304 − 965

Multiplying and dividing

Recall some of the words associated with multiplication and division.

● The product is the result of multiplying numbers.

  The **product** of 3 and 4 is 12

● The quotient is the answer to a division calculation and the remainder is anything left over.

  The **quotient** of 13 divided by 5 is 2 and the **remainder** is 3

Exercise 2.7

Use two blank dice (or make dice from nets of a cube).

Write the numbers 5, 6, 7, 8, 9 and 12 on the six faces of one die.

Write the numbers 4, 6, 7, 8, 9 and 12 on the six faces of the other die.

(a) Throw the two dice together. Write down the product of the two numbers that are rolled. (Your partner will check your answers in the next step.)

(b) After ten throws, swap with your partner and check each other's answers.

(c) Play this game as often as you can, until you are both getting all your answers right.

When you are confident that you know your times tables really well, there are some useful extensions that you can practise. The next exercise will help you to extend your times tables in various ways.

Exercise 2.8

1 Write down the answers to these doubles.

(a) 2 × 15   (c) 2 × 14   (e) 2 × 16

(b) 2 × 18   (d) 2 × 17   (f) 2 × 19

2 Write down the answers to these doubles. They use higher numbers than those in question 1

(a) 2 × 21   (c) 2 × 45   (e) 2 × 47

(b) 2 × 33   (d) 2 × 36   (f) 2 × 29

3 Write down the answers to these. The answers are all over a hundred.

(a) 54 × 2   (c) 78 × 2   (e) 2 × 95

(b) 67 × 2   (d) 2 × 86   (f) 2 × 76

4 Write down the answers to these divisions.

(a) 54 ÷ 2   (c) 72 ÷ 2   (e) 152 ÷ 2

(b) 136 ÷ 2   (d) 178 ÷ 2   (f) 184 ÷ 2

5 Write down the answers to these divisions and write any remainders as fractions.

(a) 49 ÷ 2   (c) 199 ÷ 2   (e) 155 ÷ 2

(b) 129 ÷ 2   (d) 135 ÷ 2   (f) 179 ÷ 2

6 Write down the answers to these divisions and write any remainders as fractions.

(a) 49 ÷ 3   (e) 112 ÷ 8   (i) 315 ÷ 4

(b) 89 ÷ 7   (f) 199 ÷ 12   (j) 208 ÷ 8

(c) 126 ÷ 9   (g) 246 ÷ 5   (k) 305 ÷ 7

(d) 145 ÷ 11   (h) 247 ÷ 8   (l) 351 ÷ 11

7 Write out these times tables in full:

(a) 13 times table

(c) 17 times table

(b) 15 times table

(d) 19 times table

8 Write down the answers. They have no remainders.

(a) $104 \div 13$

(e) $135 \div 15$

(i) $180 \div 15$

(b) $90 \div 15$

(f) $228 \div 19$

(j) $209 \div 19$

(c) $153 \div 17$

(g) $143 \div 13$

(k) $136 \div 17$

(d) $152 \div 19$

(h) $85 \div 17$

(l) $156 \div 13$

9 Write down the answers. They may have remainders.

(a) $95 \div 13$

(e) $156 \div 13$

(i) $120 \div 13$

(b) $168 \div 15$

(f) $199 \div 19$

(j) $129 \div 15$

(c) $105 \div 17$

(g) $210 \div 17$

(k) $120 \div 19$

(d) $180 \div 19$

(h) $133 \div 19$

(l) $120 \div 17$

⇨ Order of operations

Look at this calculation.

$15 - 4 \times 3$

You could read it as: $(15 - 4) \times 3 = 11 \times 3 = 33$

Or as: $15 - (4 \times 3) = 15 - 12 = 3$

As you can see, there are two very different answers.

The rule is that if you have **mixed operations** (that is a mix of adding, subtracting, multiplying and dividing) then you multiply or divide before you add or subtract – unless part of the calculation is in brackets and then you must do the calculation in the brackets first.

Use the expression BIDMAS to help you remember the order of operations.

B	Brackets
I	Indices
D	Divide
M	Multiply
A	Add
S	Subtract

You will look at indices in another chapter.

Examples:

Use BIDMAS to work these out.

(i) $45 - 9 \times 3$ (ii) $28 - 45 \div 9$ (iii) $45 - 7 \times 6$

(i) $45 - (9 \times 3)$ (ii) $28 - (45 \div 9)$ (iii) $45 - (7 \times 6)$
$\quad\quad = 45 - 27$ $\quad\quad\quad = 28 - 5$ $\quad\quad\quad = 45 - 42$
$\quad\quad = 18$ $\quad\quad\quad = 23$ $\quad\quad\quad = 3$

Exercise 2.9

1 Write down the answers.

(a) $7 + 9 - 3$ (d) $3 \times 5 - 6$

(b) $4 \times 5 - 6$ (e) $15 \div 5 \times 6$

(c) $3 \times 8 \div 6$ (f) $45 \div 5 \times 6$

2 Write down the answers. Remember to multiply or divide first.

(a) $3 + 4 \times 5$ (d) $17 - 45 \div 5$

(b) $12 - 36 \div 9$ (e) $12 + 120 \div 10$

(c) $5 + 14 \div 2$ (f) $15 + 5 \times 6$

3 Write down the answers. Remember to do the calculations in the brackets first.

(a) $6 - (5 - 2)$ (d) $25 - (4 \times 3) + 9$

(b) $15 + (14 - 6)$ (e) $(8 \times 9) \div (4 \times 6)$

(c) $(12 \times 3) - (7 \times 2)$ (f) $12 \times (8 + 4) \div 6$

4 Use BIDMAS to remind you how to do these calculations in the correct order.

(a) $5 + (3 + 4) \times 2$ (d) $56 \div 8 - 3 \times 2$

(b) $4 \times 5 - 3 \times 2$ (e) $55 \div (8 - 3) \times 2$

(c) $72 \div 8 - 45 \div 5$ (f) $(56 + 8) \div (4 \times 2)$

19|5

5 Try these, with larger numbers.

(a) 25 + (13 + 4) × 9

(b) 12 × 5 − 13 × 3

(c) 210 − 7 × 25 ÷ 5

(d) (210 − 7) × 25 ÷ 5

(e) 56 + (121 ÷ 11) × 2

(f) (56 + 120) ÷ 11 × 2

6 Try these, with more stages.

(a) 207 ÷ (171 ÷ 19) + (85 ÷ 17)

(b) (90 + 17 × 5) ÷ (245 ÷ 7)

(c) 144 ÷ (155 − 139) × 5

(d) (125 − 7 × 11) − 133 ÷ 7

(e) 385 ÷ (202 − 167) × (156 ÷ 13)

(f) 11 × (367 − 202) ÷ (135 ÷ 9)

⇨ Problem solving

In this next exercise you are going to use the skills you have learnt in this chapter to solve problems. If the question is a sentence your final answer should be a sentence or a short phrase. For example:

Q: What is the largest whole number that can be rounded to 100, to the nearest 10

A: The largest number is 104

Exercise 2.10

Remember to write down the calculation you are doing and then do it mentally. That way, you can check your calculation.

1 Emil has rounded a number to 600 000, to the nearest hundred thousand.

(a) What is the largest possible value of the number he started with?

(b) What is the smallest possible value of the number he started with?

Problem solving

25

2 Mia has rounded a number to 24 000 000, to the nearest million.

   (a) What is the largest possible value of the number she started with?

   (b) What is the smallest possible value of the number she started with?

3 This table shows the population of England and Wales from 1811 to 2011

Year	Population
1811	10 150 037
1861	20 066 224
1911	36 070 492
1961	43 757 888
2011	56 075 900

   (a) How much greater, to the nearest million, was the population in 2011 than in 1911?

   (b) How much greater, to the nearest hundred thousand, was the population in 2011 than in 1811?

   (c) Approximately how many times greater was the population in 2011 than in 1811?

   (d) In which fifty years did the population grow the most?

   (e) In which fifty years did the population grow the least?

4 Each pupil in a class of 19 is given 11 textbooks on the first day of term. How many textbooks is that in total?

5 The school chef orders 24 pints of milk a day on Monday, Tuesday and Thursday and 36 pints of milk on Wednesday and Friday. How many pints a week is that?

6 In a survey of our class we found the number of people in each household. These are our results.

Number in a household	Number of households
2	2
3	1
4	5
5	4
6	3

How many people is that in total?

7 In a survey of the school we found that 72 of us had one pet, 25 had two pets and 6 had three pets. How many pets in all is that?

8 The sum of two numbers is 24 and their product is 140
What are the numbers?

9 The sum of two numbers is 31 and their product is 228
What are the numbers?

10 We share out a bumper jar of 'Smarmies' among 24 of us. We each get 15 and we give the 12 left over to our teacher. How many 'Smarmies' were in the jar?

11 There were 156 strawberries to share among 18 of us. We all had an equal number but there were some left over, which we gave to our teacher.

(a) How many strawberries did we each get?

(b) How many went to the teacher?

12 For the school play we have sold 25 tickets costing £6, 75 tickets costing £8 and 20 children's tickets at £4
How much money have we taken?

3 Calculations

In Chapter 2, you concentrated on solving problems mentally, without having to write down calculations. It is useful to be able to do this but there are times when you will need **written calculations**. In this chapter you will revise written methods and solve some problems.

⇨ Addition and subtraction

One reason why the pages in mathematics exercise books are marked in squares is to help you to calculate accurately. If you think of the squares as **rows** and **columns**, then you need to take care to write the numbers in the correct columns, lining up the thousands, hundreds, tens and units correctly.

Example:

Add: 23 098 + 7435 + 353 + 54

	TTh	Th		H	T	U
	2	3		0	9	8
		7		4	3	5
				3	5	3
+					5	4
	3	0		9	4	0
	1				2	2

$8 + 5 + 3 + 4 = 20$

Note the small carried numbers here. Write them under the line to check that you have written the correct whole number. In this case it is 20

Just as for addition, when you subtract you put the numbers into a frame and calculate, column by column. You start with the units column, subtracting each time. If the top number is smaller than the bottom number then you take 1 from the column on the left and add it to the top number. This is called **changing** or **decomposition**.

Example:

Subtract: 32 145 − 17 294

	TTh	Th		H	T	U
	$^2\cancel{3}$	^{1}$^1\cancel{2}$		^{1}$^0\cancel{1}$	^{1}4	5
−	1	7		2	9	4
	1	4		8	5	1

In the **U** column, 5 − 4 = 1

In the **T** column, you cannot subtract 9 from 4 so take 1 ten from the ten (or 1) in the **H** column to leave 0. Add the 1 to the tens column to make 14

14 − 9 = 5

Continue through the **Th** and **TTh** in the same way.

Finally, use the inverse principle to check the answer mentally. Add the answer to the number that was subtracted – your answer should match the number on the top line.

	1	7		2	9	4
+	1	4		8	5	1
	3	2		1	4	5

Correct!

Example:

Subtract: 35 002 − 17 294

	TTh	Th		H	T	U
	$^2\cancel{3}$	$^{14}\cancel{5}$		$^9\cancel{0}$	$^9\cancel{0}$	$^1 2$
−	1	7		2	9	4
	1	7		7	0	8

In the **U** column you cannot take 4 from 2, but neither can you take 1 from the 0 in the **T** column or the 0 in the **H** column. The 5 in the **Th** column stands for 5 thousand so take 1 from there. 5000 = 4990 + 10 Write the 1 before the 2 in the **U** column and adjust the numbers in all the other columns.

Then subtract through the columns in the usual way.

Remember to check your answer **mentally**.

	1	7		2	9	4
+	1	7		7	0	8
	3	5		0	0	2

Correct!

You must be careful when there are zeros in the top line. For example, you cannot decompose (split) 0 tens into tens and units, so keep moving left until you find the next non-zero number.

Exercise 3.1

Calculate the answers to these additions and subtractions.

Show all your working, including the carried numbers.

1 1465 + 3421

2 4515 − 1264

3 3425 + 108 + 97

4 4302 − 1765

5 3465 + 19 + 395

6 34 172 − 6519

7 31 207 + 535 + 1499

8 19 011 − 17 099

9 14 065 + 9915 + 9

10 19 001 − 9499

11 There are 1407 pupils in the school this September. Last September there were 1368. How many more pupils are there in the school this September than last?

12 There were 4719 books in the library last summer but the librarian has added 815 more. How many books are there in the library now?

13 What is the difference between forty-five thousand, seven hundred and six and seventy-two thousand and eleven?

14 What is the sum of fifteen, six hundred and eight and twenty-six thousand and nineteen?

15 In an election, 17 803 people voted for Dr Brown, 6415 people voted for Mrs Black and 11 405 people voted for Mr Green.

(a) How many people voted?

(b) How many more votes did Dr Brown get than Mrs Black?

16 (a) There are 15 019 more people living in Puddletown than in Mudford. If the population of Puddletown is 45 195, what is the population of Mudford?

(b) The population of nearby Brownton is 20 156 less than the population of Puddleton. How much greater is the population of Mudford than the population of Brownton?

21/5

⇨ Multiplication

Short multiplication

Just as with addition and subtraction, you need to start from the units column, moving to the tens, then to the hundreds and then the thousands.

Example:

Multiply: 7126 × 8

TTh	Th		H	T	U
	7		1	2	6
				×	8
5	7		0	0	8
	1		2	4	

*Always leave room for an extra column to the left, as your answer is likely to be bigger than the original number.

6 × 8 = 48

Write 8 in the units column and 4 under the line in the tens column.

Multiplication by multiples of 10, 100, 1000

You know that when you multiply a number by 1000 the number in the units column moves along to the thousands column.

$$45 \times 1000 = 45\,000$$

Therefore, when you multiply by a multiple of 10, 100 or 1000, break the multiplication into stages.

$$106 \times 400 = 106 \times 100 \times 4$$
$$= 10\,600 \times 4$$

When you are doing written calculations, write the 0s in the frame first.

As your answer will be large, you should estimate first so you know how many columns you will need.

Example:

Multiply: 8145 × 400

Estimate: 8000 × 400 = 3 200 000 so your frame will need to include an **M** column.

Put the 0s in first.

M		HTh	TTh	Th		H	T	U
				8		1	4	5
				×		4	Ø	Ø
						0	0

To remember, say in your head: 'Cross out the naughty noughts and write them here.'

Then multiply, column by column.

M		HTh	TTh	Th		H	T	U
				8		1	4	5
				×		4	Ø	Ø
3		2	5	8		0	0	0
			1	2				

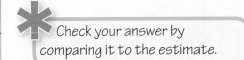

Check your answer by comparing it to the estimate.

Exercise 3.2

Calculate the answers to these. Show all your working, including the carried numbers. Remember to estimate first, so that you know how many columns to set out.

1 435 × 9

2 3126 × 7

3 436 × 60

4 345 × 400

5 1795 × 9

6 7236 × 500

7 483 × 7000

8 3465 × 12

9 6436 × 3000

10 3597 × 8000

11 Multiply six hundred and nine by twelve.

12 Multiply five thousand, six hundred and five by five hundred.

13 I bought a plant that was 412 mm high. Now it is five times as high. What is its height now?

14 An Airbus 380 seats 527 people. How many people can it carry in a week if it makes 30 flights?

15 There are 126 sugar cubes in a box. How many sugar cubes are there in 500 boxes?

16 Over a year, the school chef orders 400 packets of table napkins. If there are 144 napkins in a packet, how many napkins, in total, is that?

17 There are:

- 60 seconds in a minute

- 60 minutes in an hour

- 24 hours in a day.

(a) How many seconds are there in 24 hours?

(b) How many seconds are there in a week?

18 In the year 200, in Roman Britain, the population of London was 61 thousand. In 2000 it was 120 times bigger. What was the population of London in 2000?

Multiplication by a two-digit number

In the exercises you have completed so far, you only needed to multiply by one number. When you need to multiply by a two-digit number that is not a multiple of 10 then you need to multiply twice, once for each digit.

There are several ways of doing this.

Multiplication by factors

Example:

Multiply: 214 × 24

As   24 = 4 × 6

Then   214 × 24 = 214 × 4 × 6

             = 856 × 6

             = 5136

Th	H	T	U
	8	5	6
		×	6
5	1	3	6
	3	3	

Multiplication by splitting into tens and units (partitioning)

For this method, partition or split the smaller number into two parts – tens and units. Then multiply by each part. Finally, add the two answers together.

Example:

Multiply: 562 × 36

    562 × 36 = (562 × 30) + (562 × 6)

Estimate: 600 × 40 = 24 000, so you will need a **TTh** and a **Th** column.

TTh	Th		H	T	U
			5	6	2
		×	3	0	
1	6		8	6	0
	1				

TTh	Th		H	T	U
			5	6	2
			×		6
0	3		3	7	2
			3	1	

Add the two products.

	TTh	Th		H	T	U
	1	6		8	6	0
+		3		3	7	2
	2	0		2	3	2
	1	1		1		

Long multiplication

For long multiplication, the method is similar but you combine the three steps into one frame.

Example:

Multiply: 1859 × 56

Estimate: 2000 × 60 = 120 000

Next to each row, write down what you are multiplying by.

HTh	TTh	Th		H	T	U
		1		8	5	9
			×		5	6
	1	1_5		1_3	5_5	4
+	9_4	2_2		9_4	5	0
1	0	4		1	0	4
		1		1		

× 6   Write the carried numbers from the multiplication in these rows.

× 50

Write the carried numbers from the addition down here.

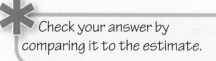
Check your answer by comparing it to the estimate.

21/5

Exercise 3.3

Calculate the answers. Show all your working, including the carried numbers. Remember to estimate first, so that you know how many columns to set out.

1 Use factors to complete these multiplications.

(a) 236 × 24 12×2 8×3 (d) 719 × 36
                  6×4

(b) 428 × 56   (e) 1324 × 48

(c) 595 × 27   (f) 2056 × 49

2 Use partitioning to complete these multiplications.

(a) 614 × 43   (d) 208 × 73

(b) 237 × 51   (e) 1536 × 47

(c) 417 × 39   (f) 2184 × 93

3 Use long multiplication to complete these multiplications.

(a) 123 × 21   (d) 675 × 91

(b) 214 × 34   (e) 756 × 85

(c) 438 × 57   (f) 946 × 38

4 Multiply seven hundred and eighty-nine by forty-seven.

5 The school needs to buy some more lockers. If one set of lockers costs £126, what is the cost of 24 sets of lockers?

6 There are 16 ounces in a pound. How many ounces are there in 735 pounds?

7 There are 2240 pounds in a ton. How many pounds are there in 37 tons?

8 Julius Caesar said with a smile: 'One seven six oh yards in a mile.'

It is 54 miles from London to Brighton. How many yards is that, if there are 1760 yards in a mile?

9 What is the area of a rectangular field that is 745 m long and 84 m wide?

10 The cost of an adventure holiday is £354 for adults and £272 for children. What is the total cost for a party of 13 adults and 45 children?

⇨ Division

When dividing, work from the highest number down to the smallest. This is instead of starting with units and then going up through tens to hundreds.

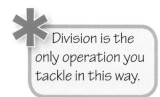
Division is the only operation you tackle in this way.

The number columns are still important in division.
Make sure that every digit in your answer is in the correct place.

Example:

Divide: $7326 \div 9$

	Th	H	T	U	
		8	1	4	
9	7	^{7}3	^{1}2	^{3}6	

You cannot divide $7 \div 9$, carry the 7

$73 \div 9 = 8$ r 1 so write 8 in the H column and carry 1

$12 \div 9 = 1$ r 3 so write 1 in the T column and carry 3

$36 \div 9 = 4$ so write 4 in the U column.

Sometimes a division will have a remainder.

Example:

Divide: $6341 \div 6$

	Th	H	T	U	
		1	0	5	6	r 5
6	6	3	^{3}4	^{4}1	

$6341 \div 6 = 1056$ r $5 = 1056 \frac{5}{6}$

Dividing by factors

It is sometimes easier to complete a division in two stages, by **dividing by factors**.

Example:

Divide: $348 \div 48$

$$348 \div 48 = 348 \div 6 \div 8$$
$$= 58 \div 8$$
$$= 7 \text{ r } 2 = 7\frac{1}{4}$$

Dealing with a remainder

When you are solving problems, you often have to decide what to do with the remainder.

- Sometimes you may just have something left over.

- Sometimes you might want to write it as a fraction in its lowest terms.

- Sometimes you may need to round up or round down.

Think about this question.

6341 people are divided into groups of six. How many groups are there?

$$6341 \div 6 = 1056 \text{ r } 5$$

Each person must be in a group, so you will have 1056 full groups and one smaller group. You cannot just ignore five people. You will have to **round up**. Therefore there will be 1057 groups.

Now think about this one.

A farmers' cooperative has 6341 eggs. An egg box takes six eggs. How many boxes can the cooperative fill?

$$6341 \div 6 = 1056 \text{ r } 5$$

They can fill 1056 egg boxes but the remainder will not fill another box. They can only sell full boxes, so they have to **round down**. Therefore they will have 1056 full boxes.

Calculate the answers to these. Show all your working, including the carried numbers. If there are remainders, write them as fractions in their lowest terms.

1 $732 \div 4$

2 $409 \div 6$

3 $742 \div 3$

4 $935 \div 5$

5 $5243 \div 4$

6 $8136 \div 7$

7 $8436 \div 9$

8 $2465 \div 11$

9 $8177 \div 8$

10 $2435 \div 12$

Answer the next set of questions by dividing by factors. Write any remainders as fractions in their lowest terms.

11 $192 \div 24$

12 $324 \div 36$

13 $756 \div 63$

14 $825 \div 55$

15 $3652 \div 44$

16 $3378 \div 27$

17 $8974 \div 49$

18 $3536 \div 72$

19 $7616 \div 28$

20 $8776 \div 32$

21 Divide nine hundred and twelve by seven.

22 Divide six thousand and eleven by eight.

23 1060 children each eat a slice of pizza. One slice is an eighth of a whole pizza. How many pizzas do they eat?

24 There are 8 pints in a gallon. The school cook orders 2453 pints of milk per week. How many gallons is that?

25 There are 12 inches in a foot. How many feet are there in 1144 inches?

26 The volume of a cuboid is 864 mm³. If it is 12 mm long and 9 mm wide, what is its height?

27 The school pond holds 2354 litres of water and a bucket can take 3 litres. How many full buckets of water must be removed, to empty the pond?

Division by multiples of 10

Consider the calculation:

$$450 \div 30$$

You know that you can divide a number into its factors.

$30 = 10 \times 3$ so the calculation becomes:

$$450 \div 30 = 450 \div 10 \div 3$$
$$= 45 \div 3$$
$$= 15$$

You can take out the first stage of this calculation and just write:

$$450 \div 30 = 45 \div 3$$

Or think of it as:

$$45\cancel{0} \div 3\cancel{0} = 15$$

* Only division can be tackled like this.

Examples:

Divide: (i) $640 \div 40$ (ii) $4000 \div 800$ (iii) $240\,000 \div 6000$

(i) $640 \div 40 = 64\cancel{0} \div 4\cancel{0}$
$$= 16$$

(ii) $4000 \div 800 = 40\cancel{0}\cancel{0} \div 8\cancel{0}\cancel{0}$
$$= 5$$

(iii) $240\,000 \div 6000 = 240\,\cancel{0}\cancel{0}\cancel{0} \div 6\cancel{0}\cancel{0}\cancel{0}$
$$= 40$$

Exercise 3.5

Calculate the answers.

1 $700 \div 20$

2 $400 \div 50$

3 $2700 \div 30$

4 $8400 \div 400$

5 $2700 \div 900$

6 $34\,000 \div 200$

7 $56\,000 \div 7000$

8 $80\,000 \div 4000$

9 $360\,000 \div 9000$

10 $500\,000 \div 2000$

11 450 000 ÷ 50   16 300 000 ÷ 600

12 40 000 ÷ 800   17 440 000 ÷ 11 000

13 360 000 ÷ 120   18 960 000 ÷ 1200

14 500 000 ÷ 400   19 500 000 ÷ 800

15 54 000 ÷ 900   20 700 000 ÷ 40

Long division

Before you think about long division, remember what happens in simple division.

Example:

Think back to dividing 7326 by 9

	Th	H	T	U	
		8	1	4	
9	7	^{7}3	^{1}2	^{3}6	

You cannot divide 7 ÷ 9, carry the 7

73 ÷ 9 = 8 r 1 so write 8 in the H column and carry 1

12 ÷ 9 = 1 r 3 so write 1 in the T column and carry 3

36 ÷ 9 = 4 so write 4 in the U column.

To get the first remainder of 1, the procedure was:

73 **divided** by 9 is 8 because

9 **multiplied** by 8 is 72

73 **subtract** 72 is 1

You can continue to calculate remainders in this way and could write out the calculation like this.

Example:

$7326 \div 9$

	Th	H	T	U	
		8	1	4	
9	7	3	2	6	Divide: $73 \div 9 = 8$
	7	2			Multiply: $9 \times 8 = 72$
		1	2		Subtract: $73 - 72 = 1$ and pull down the 2
			9		Divide: $12 \div 9 = 1$ and multiply: $9 \times 1 = 9$
			3	6	Subtract: $12 - 9 = 3$ and pull down the 6
			3	6	Divide: $36 \div 9 = 4$ and multiply $9 \times 4 = 36$
			–	–	Subtract: $36 - 36 = 0$

This is called long division because of the long tail. Look at the instructions down the right-hand side.

Divide, multiply, subtract, pull down the number, divide, multiply, subtract, pull down the number, . . .

Stick to that rhythm as you work through your division.

When dividing by two-digit numbers, use the same steps. Because you will not know the times tables for all these two-digit numbers, you need to **estimate** each step of the division and then **check by multiplying**.

Consider dividing 7546 by 24

The first step is 75 ÷ 24

Estimate: 70 ÷ 20 = 3.5

Check by multiplying by 3

$24 \times 3 = 72$

Write all this in a frame, with the multiplications to the side. You can put your working commentary in the frame as well.

<table>
<tr><td></td><td></td><td>Th</td><td>H</td><td>T</td><td>U</td></tr>
<tr><td></td><td></td><td></td><td>3</td><td>1</td><td>4</td></tr>
</table>

*Round 75 down to 70 when estimating, because you want the answer to be **less than** 75*

Example:

Divide: 7546 ÷ 24

		Th	H	T	U	
			3	1	4	
2	4	7	5	4	6	Divide.
		7	2			Multiply.
			3	4		Subtract and pull down.
			2	4		Divide, multiply.
			1	0	6	Subtract and pull down.
				9	6	Divide, multiply.
				1	0	Subtract.

2	4
×	3
7	2

2	4
×	4
9	6

7546 ÷ 24 = 314 r 10

Check your answer by an estimated multiplication

$25 \times 300 \approx 7500$

(25 is a better estimate of 24 than 20)

Exercise 3.6

Use long division to work out the answers. The first 15 questions have no remainders but the rest may have remainders.

1	576 ÷ 4	6	1716 ÷ 12
2	774 ÷ 3	7	665 ÷ 19
3	1645 ÷ 5	8	714 ÷ 17
4	2088 ÷ 6	9	923 ÷ 13
5	2394 ÷ 7	10	690 ÷ 15
11	864 ÷ 24	16	890 ÷ 37
12	828 ÷ 36	17	990 ÷ 64
13	945 ÷ 35	18	800 ÷ 53
14	943 ÷ 23	19	850 ÷ 34
15	918 ÷ 54	20	904 ÷ 72
21	1998 ÷ 27	26	7243 ÷ 38
22	4998 ÷ 34	27	4562 ÷ 31
23	3529 ÷ 63	28	7448 ÷ 49
24	3030 ÷ 42	29	4908 ÷ 35
25	2548 ÷ 26	30	9324 ÷ 74

⇨ Problem solving

In this next exercise you are going to use the skills you have learnt in this chapter to solve problems. You will need to use **long division** to answer some of the questions.

Exercise 3.7

Answer these questions. If there is a **remainder**, decide whether you should round the answer up or down. Alternatively, you might want to give your answer as a **fraction** or **decimal**.

1 Divide four hundred and fifty-six by twenty-four.

2 Divide six thousand and twelve by thirty-six.

3 The average life of a halogen light bulb is 2000 hours. How many days is this?

4 There are 14 pounds in a stone. How much does a 215-pound man weigh, in stones and pounds?

5 Each railway carriage has 74 seats. How many carriages will be needed for 4078 people to have seats?

6 The area of a rectangle is 1026 cm². If the length of the rectangle is 38 cm, what is the width?

7 A factory makes 7000 Christmas decorations a day. If the decorations are packed in boxes of 37, how many full boxes are packed each day?

8 We have to put out chairs for the audience at the school play. There are 17 chairs in a row and we need to seat 205 people. How many rows is that?

9 We spent £55 on bricks to repair a wall. If bricks cost 44p each, how many bricks did we buy?

10 One English pound is worth 55 Russian roubles. What is the value of 10 000 roubles, in pounds?

Exercise 3.8: Summary exercise

Calculate the answers to these. Show all your working, including the carried numbers.

1 $1987 + 9 + 495$

2 $1036 - 549$

3 $432 \div 9$

4 346×7

5 543×23

6 $12\,000 \div 500$

7 $986 \div 34$

8 $45\,345 + 675 + 7346$

9 $12\,435 - 3678$

10 $6438 \div 37$

11 A train left London with 542 passengers. At Crewe, 126 joined the train and 94 got off. How many passengers were there on the train when it left Crewe?

12 A field measures 45 m by 79 m. A farmer buys fertiliser for the field, at a cost of 59p per square metre. What does it cost to fertilise the field?

13 I am flying to New Zealand via Los Angeles. It takes 10 hours 10 minutes to fly from London to Los Angeles and 12 hours 40 minutes from Los Angeles to New Zealand. How many minutes is that altogether?

14 The school bought 24 globes costing £27 each and 36 compasses costing £14 each. What did they spend altogether?

15 The school buys a pack of 24 battery calculators costing £300 altogether and then 36 solar-powered calculators costing £14 each.

   (a) Which calculator was better value and by how much?

   (b) What did the school pay in total?

16 In a competition, Ali won £12 456 and Akram won £5672

   (a) What did they win in total?

   (b) How much more did Ali win than Akram?

17 Fred earned £1725 by working for 23 days and Freda earned £1444 by working for 19 days. Who was paid more per day?

18 A dogs' home orders 6000 cans of dog food a month. They calculate each dog eats 26 cans per month. How many dogs can they feed?

19 My back garden measures 32 m by 15 m. Next to the house is a patio measuring 15 m by 5 m. The remainder of the garden is grass. What area of my garden is grass?

20 Cuboid A is measures 17 cm by 24 cm by 9 cm and Cuboid B measures 25 cm by 11 cm by 13 cm.

   (a) Which cuboid has the greater volume?

   (b) How much greater is the volume of this cuboid?

Carefully copy these squares on to a piece of centimetre-squared paper. The sizes must be exactly as shown.

| 1 |

| 1 |

| 2 |

| 3 |

| 5 |

| 8 |

13

Cut them out and see if you can arrange them so they fit inside a rectangle 13 cm by 21 cm.

What do you notice about the pattern of rectangles that you have made?

Write down the lengths of the sides of the squares. Arrange them in order, from smallest to largest. Try to work out the rule and continue the sequence.

Write the rule like this:

If the first number is a,

and the second number is b,

then the following number will be ...

This sequence of numbers is called the **Fibonacci sequence**, named after the mathematician Leonardo of Pisa, the son of Bonacci and most commonly known as Fibonacci.

His sequence is remarkable because the numbers are found so often in nature. For example, look at your hands. You have 5 fingers. Each finger has 2 knuckles and is divided into 3 sections. 1, 2, 3 and 5 are all Fibonacci numbers.

Do some more research into Fibonacci and his numbers and share what you find with the class.

Angles and triangles

11/5/2020

⇨ Angles

You know that an **angle** is formed at the point where two lines meet.

One line

Two lines forming two pairs of equal angles

Angles are measured in **degrees**. One full turn is equal to 360°

This diagram shows an angle of 144°, measured with a full circle protractor.

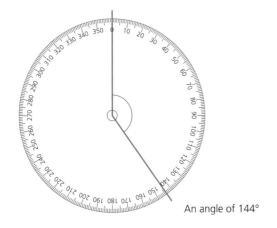

An angle of 144°

Types of angle

Before you measure an angle, you should consider whether it is acute, obtuse, right or reflex.

Right angle	Acute angle	Obtuse angle	Reflex angle
Exactly 90°	Less than 90°	Between 90° and 180°	Between 180° and 360°

> ✳ You can identify the angle you are working with by drawing a curved line, or **arc**, around it. A **right angle** is marked by a special small square.

Measuring angles

You need to use a **protractor** to measure the size of an angle. Note that the protractor has two scales, one measuring clockwise from zero and the other anticlockwise from zero.

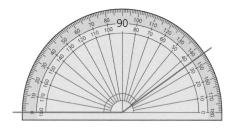

Is the angle being measured 35° or 145°?

Look carefully at the two arms of the angle. It is an obtuse angle and is 145°

To measure a reflex angle with a half circle protractor, you need to measure the other angle formed by the arms of the reflex angle and subtract it from 360°, the full turn.

Make an estimate first.

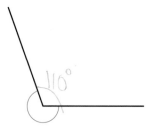

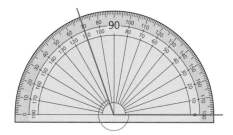

Reflex angle

Approximate size 240°

The obtuse angle measures 110°

The reflex angle measures 360° − 110° = 250°

When measuring angles, follow these steps.

90-180   180-360

- Decide whether the angle is acute, obtuse or reflex.

- Estimate the size of the angle.

- Decide which arm of the angle to line up with 0° on the protractor.

- Make sure you use the correct scale (inside or outside).

- To measure a reflex angle, measure the obtuse or acute angle and subtract it from 360°

11/5/2020

Exercise 4.1

For each angle:

(a) Write down whether it is a right angle, an acute, an obtuse or a reflex angle.

(b) Estimate the size of the angle.

(c) Measure the angle.

1

acute 60°

4

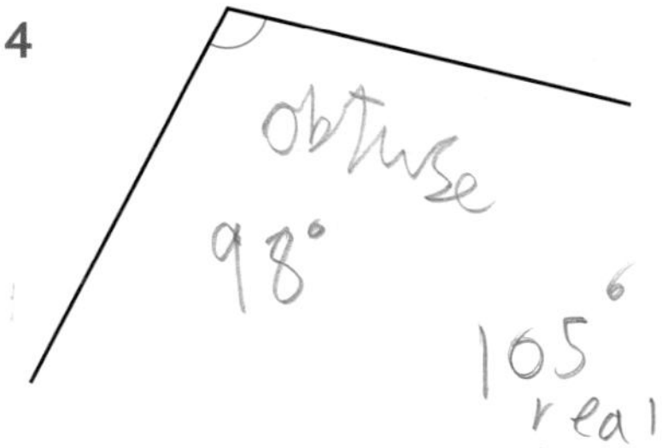

obtuse
98°
105° real

2

Right
90°

5

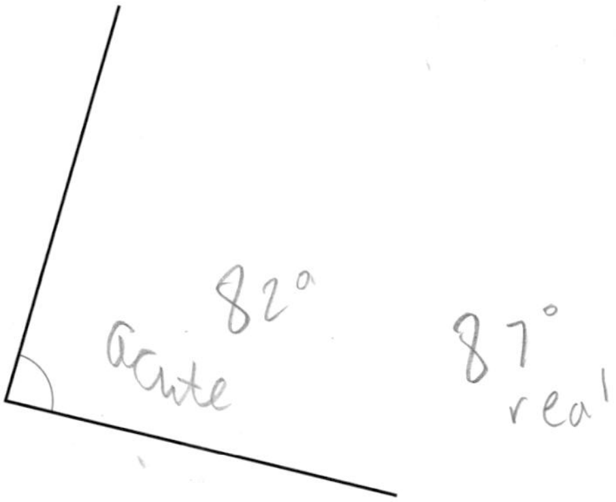

82°
acute
87° real

3

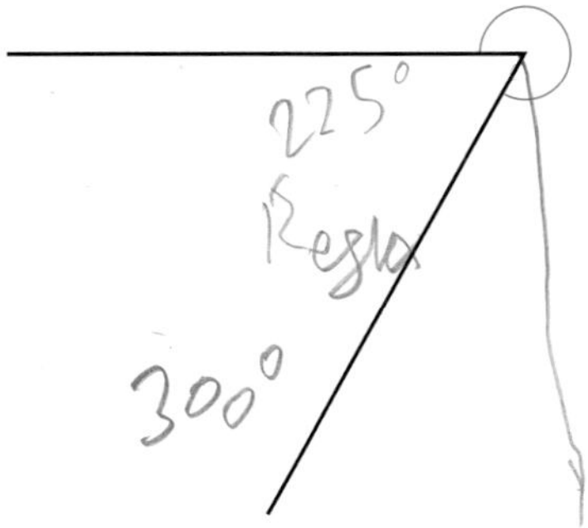

225°
Refla
300°

6

100°
obtuse
113° real

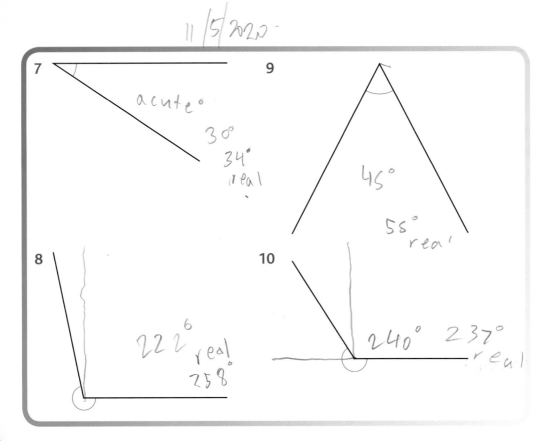

11/5/2020

7

acute°

30°
34°
real

9

45°

55°
real

8

222⁶ real
258°

10

240° 237°
real

⇨ Triangles

Angles in a triangle

A **triangle** is a two-dimensional (**2D**) shape with three sides.

A triangle has three corners (or points or **vertices**).

At each corner there is an **angle**.

You can use the letters that label the vertices to describe triangles and angles.

This triangle is $\triangle ABC$.

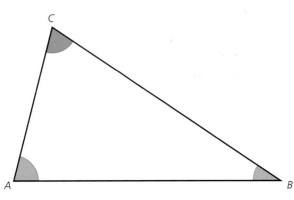

The symbol △ means 'triangle' The symbol ∠ means 'angle'.

The blue angle is called ∠*A* or ∠*CAB*, as it is at the point where sides *CA* and *AB* meet.

The green angle is ∠*B* or ∠*ABC* as it is at the point where sides *AB* and *BC* meet.

The red angle is ∠*C* or ∠*BCA* as it is at the point where sides *BC* and *CA* meet.

Copy the triangle above, colouring the three angles red, green and blue. Cut it out. Then put the angles together, like this.

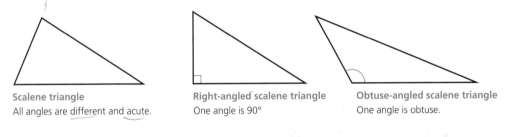

The three angles together form a **straight line**, which is an angle of 180°

Types of triangle

The angles in a triangle tell you which type of triangle it is.

Scalene triangle
All angles are different and acute.

Right-angled scalene triangle
One angle is 90°

Obtuse-angled scalene triangle
One angle is obtuse.

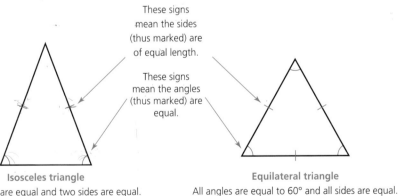

These signs mean the sides (thus marked) are of equal length.

These signs mean the angles (thus marked) are equal.

Isosceles triangle
Two angles are equal and two sides are equal.

Equilateral triangle
All angles are equal to 60° and all sides are equal.

Drawing triangles

Every triangle has three angles and three sides.

You do not need to know the size of all the angles and the lengths of all the sides to draw a triangle.

You only need to know one of these sets of three facts about the triangle:

- the lengths of three sides

- the lengths of two sides and the size of the enclosed angle

- the sizes of two angles and the length of the enclosed side

You also need a ruler, a pair of compasses, a protractor and, very importantly, a sharp, hard pencil.

Given the lengths of three sides

Example:

Draw $\triangle ABC$ in which $AB = 5$ cm, $BC = 6$ cm and $AC = 4.5$ cm.

Step 1: Draw the longest side, which is BC, accurately.

B ─────────────────────────── C

Step 2: Open your compasses to a **radius** of 5 cm (because the length of AB is 5 cm). Put the compass point on point B and draw an arc above line CB, as shown.

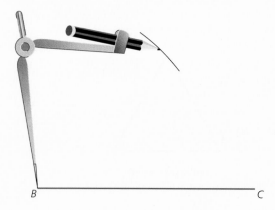

Step 3: Open your compasses to a radius of 4.5 cm (because the length of *AC* is 4.5 cm). Put the compass point on point *C* and draw an arc above line *CB*, to cross the first arc as shown. Mark the point where the arcs cross as *A*.

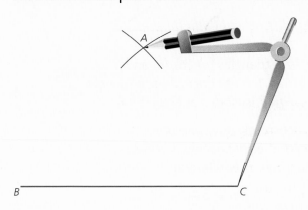

Step 4: Join *B* to *A* and *C* to *A* and neatly label your triangle.

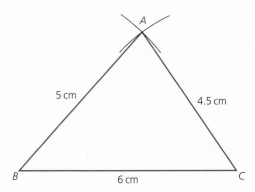

Exercise 4.2

Make sure that you have a sharp pencil, a ruler and a pair of compasses for this exercise.

1 Draw and label these triangles accurately.

(a)

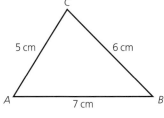

(b)

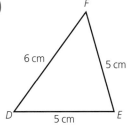

(c)

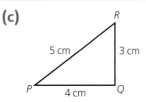

(d)

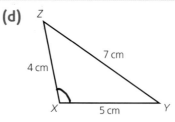

2 Measure and record all the angles in triangles (a), (b), (c) and (d) that you have drawn for question 1. Write down if each triangle is scalene, right-angled, obtuse-angled, isosceles or equilateral.

3 Draw and label these triangles accurately. Measure and record all their angles. Write down if each triangle is scalene, right-angled, obtuse-angled, isosceles or equilateral.

(a) $\triangle ABC$ in which $AB = 7$ cm, $BC = 4.5$ cm and $AC = 5.5$ cm

(b) $\triangle DEF$ in which $DE = 8.2$ cm, $EF = 5.5$ cm and $DF = 5.5$ cm

(c) $\triangle JKL$ in which $JK = 6$ cm, $KL = 6$ cm and $JL = 6$ cm

(d) $\triangle PQR$ in which $PQ = 6$ cm, $QR = 4.5$ cm and $PR = 7.5$ cm

4 Sam tried to draw a triangle with sides 9 cm, 5.5 cm and 3 cm. What do you think happened?

Given the lengths of two sides and the enclosed angle

Example:

Draw $\triangle ABC$ in which $AB = 7$ cm, $AC = 6$ cm and $\angle BAC = 53°$

Step 1: Draw a sketch of the triangle and label it. Do this to make sure that you draw the correct angle.

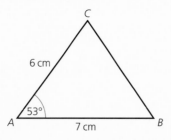

Step 2: Draw the side AB accurately.

Step 3: Use your protractor to measure and mark ∠*BAC*. Draw a line through the point *A* and the mark.

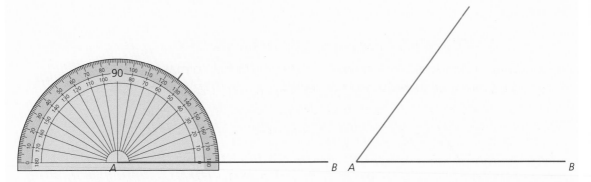

Step 4: Use your ruler to measure and mark the length *AC* and label point *C*.

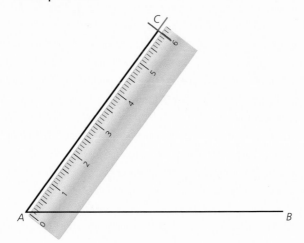

Step 5: Join *B* to *C* and label your triangle.

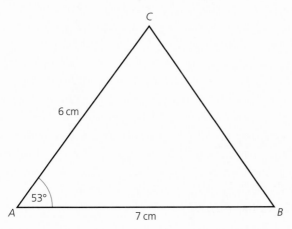

Given the sizes of two angles and the length of the enclosed side

Example:

Draw △ABC in which AB = 6.5 cm, ∠BAC = 55° and ∠ABC = 45°

Step 1: Draw a sketch of the triangle and label it. Do this to make sure that you draw the correct angles.

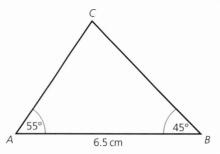

Step 2: Draw the side AB accurately.

A ———————————————— B

Step 3: Use your protractor to measure and mark ∠BAC. Draw a line through the point A and the mark.

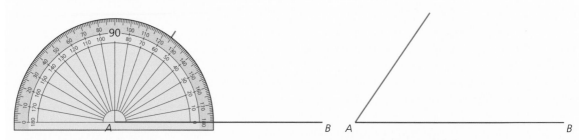

Step 4: Use your protractor to measure and mark ∠ABC. Draw a line through the point B and the mark. Label the point where the lines cross as C.

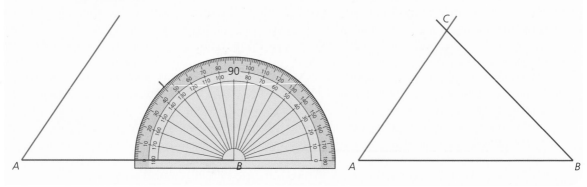

Step 5: Label your triangle.

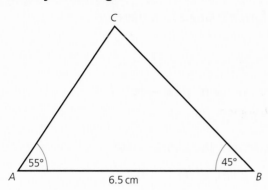

1 Draw and label these triangles accurately.

(a)

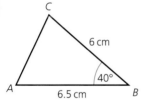

(c)

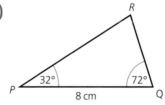

(b)

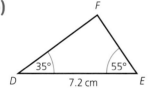

(d)

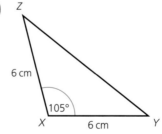

2 Measure and record all the angles in triangles (a), (b), (c) and (d) that you have drawn for question 1. Write down if each triangle is scalene, right-angled, obtuse-angled, isosceles or equilateral.

3 Draw and label these triangles accurately. Measure and record all their angles. Write down if each triangle is scalene, right-angled, obtuse-angled, isosceles or equilateral.

   (a) $\triangle ABC$ in which $AB = 80$ mm, $AC = 65$ mm and $\angle BAC = 63°$

   (b) $\triangle DEF$ in which $DE = 7$ cm, $\angle DEF = 48°$ and $\angle FDE = 66°$

   (c) $\triangle JKL$ in which $JK = 72$ mm, $KL = 72$ mm and $\angle JKL = 60°$

   (d) $\triangle PQR$ in which $PQ = 9$ cm, $\angle PQR = 44°$ and $\angle RPQ = 44°$

4 Susi tried to draw $\triangle ABC$ in which $AB = 8$ cm, $\angle CAB = 135°$ and $\angle ABC = 45°$. What do you think happened?

12/5/20 20

⇨ Calculating angles

You know that one full turn is equal to 360°
This is the angle at the centre of a circle.

A clockface is an example of a full circle divided into 12 equal sections.

Exercise 4.4

12/5

Use the fact that there are 360° degrees in a full turn to answer these questions.

1 Divide the clock into four equal parts.

$90°$

Calculate the sizes of the angles at the centre.

2 Divide the clock into six equal parts.

$60°$

Calculate the sizes of the angles at the centre.

3 Divide the clock into eight equal parts.

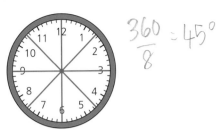

$$\frac{360}{8} = 45°$$

Calculate the sizes of the angles at the centre.

4 Divide the clock into twelve equal parts.

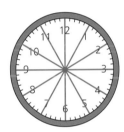

$$\frac{360}{12} = 30°$$

Calculate the sizes of the angles at the centre.

5 What is the angle between the hands of a clock at:

(a) 2 o'clock

60°

(b) 6 o'clock 180°

(c) 7 o'clock 210°

(d) 11 o'clock? 330°

6 Through what angle will the hour hand move in half an hour? 180°

7 What is the angle between the hands of a clock at:

(a) half past one 135°

15

30

45

(b) half past 12 165°

(c) half past 9 90 + 15 = 105°

(d) half past 10? 135°

8 In a quarter of an hour, through what angle will:

(a) the hour hand move

(b) the minute hand move?

HW

12/5

9 What is the angle between the hands of a clock at:

(a) quarter past 1 8 × 6° = 48°

1 min $\frac{360}{60} = 6$

1 min = 6°

(b) quarter to 10 4 × 6 = 24°

(c) quarter past 3 0°

(d) quarter to 12? 14 × 6 = 84°

10 What is the angle between the hands of a clock at:

(a) twenty past 4 $0°$

(b) ten to 6

(c) twenty to 2

(d) twenty-five to 9

(e) five past 8

(f) twenty-five past 4?

➡ **More angle facts**

12/5/2020

You know several facts about angles. You can use these to find missing angles.

Angles on a straight line add up to 180° ✓

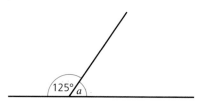

To find the value of a, subtract the given angle from 180°

$a = 180° - 125°$ Angles on a straight line

$= 55°$

Angles at a point add up to 360°

To find the value of $2 \times b$ add 120° to 90° and subtract from 360°

$2 \times b + 90° + 120° = 360°$ Angles at a point

$2 \times b = 360° - (90° + 120°)$

$= 360° - 210°$

$= 150°$

So $b = 75°$

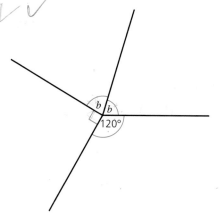

4 Angles and triangles

64

Vertically opposite angles are equal

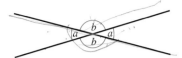

You can see that $a + b = 180°$, so it follows that the vertically opposite angles are equal.

As intersecting lines form pairs of equal angles:

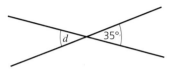

$d = 35°$ Vertically opposite angles

Angles in a triangle add up to 180°

To find the value of $2 \times c$ subtract 24° from 180°

$2 \times c + 24° = 180°$ Angles in a triangle

$2 \times c = 180° - 24°$

$= 156°$

So $c = 78°$

Using angle facts

You can combine the facts that you know, to find the sizes of missing angles.

It is important to give reasons for each statement you make about angles. This will help you with the reasoning, and allow someone else to follow your thinking.

Example:

Find the values of b, c and d.

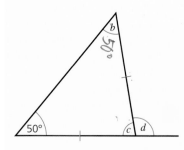

$b = 50°$ Base angles of an (isosceles) triangle

$50° + 50° + c = 180°$ Angles in a triangle

$100° + c = 180°$

$c = 80°$ ✓

$80° + d = 180°$ Angles on a straight line

$d = 100°$ ✓

12/5/2020

Exercise 4.5

Calculate the sizes of the angles marked with small letters.
Give reasons for any angle statements. *AB* and *CD* are always
straight lines.

1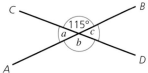

3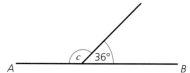

2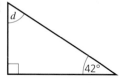

4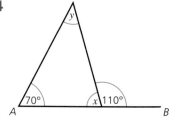

12/5/2020

5

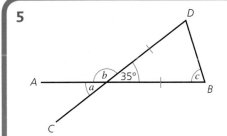

9

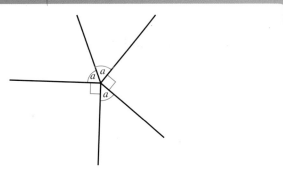

6

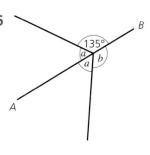

10

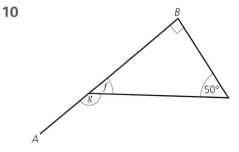

7

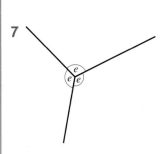

11

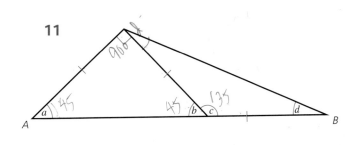

8

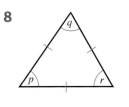

12

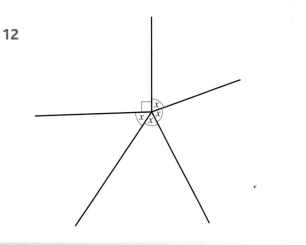

Exercise 4.6: Summary exercise

1 Measure these angles. Say whether each is a right, acute, obtuse or reflex angle.

(a)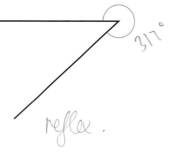

317°

reflex.

(c)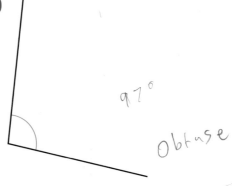

97°

obtuse

(b)

90°

Right

(d)

acute

82°

2 (a) Draw these triangles accurately.

(i)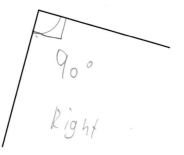

C

2.5 cm   6 cm

A   6.5 cm   B

scalene

(ii)

F

44°   68°

D   7 cm   E

isosceles

(iii)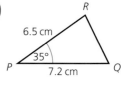

R

6.5 cm

35°

P   7.2 cm   Q

(b) Measure and write down the sizes of all the unknown sides and angles.

(c) State whether each triangle is scalene, right-angled, obtuse-angled, isosceles or equilateral.

3 Draw and label each of these triangles accurately. Measure and label all the angles. Write down if each triangle is scalene, right-angled, obtuse-angled, isosceles or equilateral.

(a) △ABC in which AB = 55 mm, BC = 45 mm and AC = 70 mm

(b) △DEF in which DE = 6.5 cm, ∠DEF = 38° and ∠FDE = 52°

(c) △JKL in which JK = 82 mm, KL = 64 mm and ∠JKL = 25°

4 What is the angle between the hands of a clock at:

(a) 8 o'clock 120°

(c) quarter to 6 91°

(e) five past 9 120°

(b) half past 9 105°

(d) twenty-five to 11 104°
10:35

(f) twenty-five past 6?

5 Calculate the sizes of the angles marked with small letters. Give reasons for any angle statements. AB and CD are always straight lines.

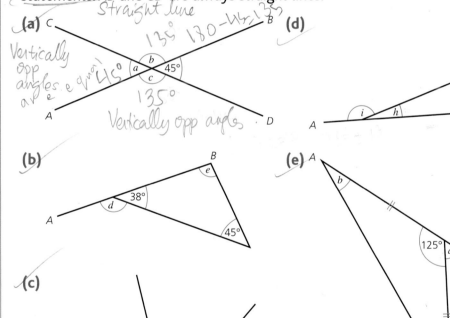

(a) Straight line
135° 180−44=135
Vertically opp angles e qual 45°
135°
Vertically opp angles

(b)

(c)

(d)

(e)

Activity – Compass patterns: the seed of life

Use your compasses to draw a circle of radius 4 cm.

Without changing the radius, put the compass point on the circumference and draw two marks on the circumference, one on each side of the first point.

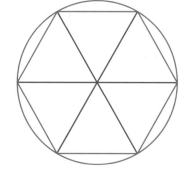

Now put the point on each of your marks and make two more marks. Do this one more time. Both marks should be on the same point, this time.

You will then have five marks and the original tiny hole where you first put your compass point.

If you join these marks, in order, you will have drawn a **regular hexagon**.

Now join each point of the hexagon to the centre of the circle. This should give you six angles of 60°. Check with your protractor.

Now draw another circle.

This time instead of drawing small arcs, draw six full circles. You should end up with this pattern, which is called the **seed of life**.

Can you expand your pattern with another set of interesting circles?

5 Number puzzles

You have done a lot of calculations and now you are going to use what you have learnt to solve some puzzles.

⇨ Missing numbers

In the next exercise, there are some calculations just like the ones you have been doing in previous chapters – but some of the **numbers** are missing.

Exercise 5.1

Copy these calculations and fill in the missing numbers.

1.
```
     □  4  5
  +     8  □  6
  ─────────────
     1  1  0  □
```

2.
```
     3  □  □  5
  −        7  3  □
  ────────────────
     2  7  3  1
```

3.
```
        □  5  □
     ×        4
  ─────────────
     1  8  1  2
```

4.
```
          1  □  3
     8 │ □  8  □
```

5.
```
        4  □  7
        7  3  □
  +     □  5  9
  ─────────────
     □  0  8  2
```

6.
```
     8  □  0  □
  −  6  4  □  8
  ─────────────
     1  5  7  5
```

7
```
        5 □
    ×   4 7
    ─────────
      □ 8 5
    2 2 0 □
    ─────────
    □ 5 □ 5
```

8
```
            □ 4
    1 7 │ 4 □ 8
          □ 4
          ───
            6 □
          □ □
          ───
          -  -
```

⇨ Missing operations

In Chapter 2 you learnt about the order of operations, which is summarised as BIDMAS.

In the next exercise, some of the **operations** are missing. You have to work them out by looking at the numbers.

Examples:

Fill in the correct operation, +, −, × or ÷, to make each calculation correct.

(i) $9 \,\square\, 6 = 15$ (ii) $4 \,\square\, 6 = 8 \,\square\, 3$ (iii) $(8 \,\square\, 6) \,\square\, 12 = 36$

(i) $9 + 6 = 15$ 15 is larger than 9 or 6 so consider + or ×

(ii) $4 \times 6 = 8 \times 3$ You have two calculations that have the same answer.

(iii) $(8 \times 6) - 12 = 36$ Make sure that you use the brackets correctly.

Exercise 5.2

Fill in the correct operation, +, −, × or ÷, to make each calculation correct.

1 $15 \,\square\, 7 = 8$ 6 $9 \,\square\, 5 \,\square\, 3 = 7$

2 $36 \,\square\, 9 = 4$ 7 $6 \,\square\, 9 \,\square\, 5 = 3$

3 $8 = 26 \,\square\, 18$ 8 $3 \,\square\, 8 \,\square\, 6 = 4$

4 $24 = 8 \,\square\, 3$ 9 $12 \,\square\, 3 = 9 \,\square\, 4$

5 $7 \,\square\, 6 \,\square\, 3 = 10$ 10 $8 \,\square\, 4 = 36 \,\square\, 3$

11 9 ☐ 3 = 24 ☐ 3   15 16 ☐ 8 ☐ 2 = 4

12 8 ☐ 6 = 12 ☐ 6   16 8 ☐ 6 = 8 ☐ (12 ☐ 3)

13 (5 ☐ 4) ☐ 2 = 2   17 12 ☐ (8 ☐ 2) = 9 ☐ (16 ☐ 2)

14 12 ☐ (8 ☐ 2) = 2   18 (6 ☐ 3) ☐ 4 = 8 ☐ (24 ☐ 6)

⇨ Using symbols

In the questions above you had some numbers, but sometimes you may only have an answer. You are used to questions such as:

If the cost of three buns is 96p, what is the cost of one bun?

but the question could look like this:

● + ● + ● = 96

You still solve this in the same way, by dividing by 3, so the answer is:

● = 32

Example:

Find the value of ●, ■ and ▲ in these three statements.

● + ● = 24

● + ■ = 20

● + ■ + ▲ = 26

● = 24 ÷ 2

So ● = 12

Then 12 + ■ = 20

■ = 20 − 12

So ■ = 8

And 12 + 8 + ▲ = 26

So ▲ = 6

Find the value of the symbols in each of these questions.

1 ● + ● = 18

● + ■ = 12

2 ◆ + ◆ + ◆ = 18

◆ + ■ = 9

3 ● + ● + ● = 24

● + ■ = 12

● + ■ + ◆ = 21

4 ● + ● + ● = 9

● + ■ + ■ = 15

● + ■ = 20 − ◆

5 ❖ + ❖ + ❖ = 20 − ✽

✽ + ★ = (❖ + ❖) × 3

✽ + ★ = 24

6 ★ + ✽ + ✽ = 13

★ − ✽ = 4

❖ + ❖ + ❖ + ❖ + ❖ = 2 × (★ + ✽)

7 ● + ● = ■

● × ● = ■

● + ■ + ◆ = 11

8 ● × ● = ●

● + ■ = ■ + ◆

● + ■ + ◆ = 10

For this set of questions there are no symbols but just words.

Use exactly the same methods as before, to solve these.

9 The sum of two numbers is 13 and their difference is 3
What are the numbers?

10 A silver medal weighs twice as much as a bronze medal. A bronze medal and two silver medals weigh the same as a gold medal. How many bronze medals weigh the same as a gold medal?

11 If two cakes cost 90p and one cake and one bun costs 75p, what is the cost of three buns?

12 A box of paints and a drawing book costs £3.50. A paintbrush and a drawing book costs £1.50. Three paintbrushes cost 72p. How much does a box of paints cost?

13 I can buy three apples and two bananas for 85p, or two apples and one banana for 50p. What is the cost of:

(a) one apple and one banana

(b) two apples?

14 Two widgets and one fidget cost £11 but one widget and two fidgets cost £13
What is the cost of one widget and one fidget?

15 Look at this menu.

1 egg and tomato	80p
1 egg and chips	£1.10
1 egg, chips and tomato	£1.40

What would be the cost of two eggs, chips and tomato?

16 I can buy one pen and two pencils for £1.50, or two pencils and one eraser for 70p or one pen, one pencil and one eraser for £1.45
I decide to buy one pen and one pencil. How much will that cost me?

⇨ Function machines

Imagine that you put a number into a machine and another number comes out. Look at this machine.

```
5 →  ┌──────┐ → 7
1 →  │  +2  │ → 3
3 →  └──────┘ → 5
```

You can see that each of the numbers that comes out is 2 more than the number that went in, as the machine is a 'plus 2' machine.

This is a 'multiply by 3' machine. You can see what comes out – but not what went in.

```
a →  ┌──────┐ → 9
b →  │  ×3  │ → 12
c →  └──────┘ → 15
```

As $3 \times a = 9$, then $a = 3$

As $3 \times b = 12$, then $b = 4$

As $3 \times c = 15$, then $c = 5$

Exercise 5.4

Calculate the value of the missing numbers that come in or out of these function machines.

1 $5 \rightarrow$ [ ] $\rightarrow a$
   $1 \rightarrow$ [$+4$] $\rightarrow b$
   $c \rightarrow$ [ ] $\rightarrow 12$

3 $5 \rightarrow$ [ ] $\rightarrow a$
   $1 \rightarrow$ [$\times 2$] $\rightarrow b$
   $c \rightarrow$ [ ] $\rightarrow 8$

2 $5 \rightarrow$ [ ] $\rightarrow a$
   $8 \rightarrow$ [-3] $\rightarrow b$
   $c \rightarrow$ [ ] $\rightarrow 14$

4 $9 \rightarrow$ [ ] $\rightarrow a$
   $18 \rightarrow$ [$\div 3$] $\rightarrow b$
   $c \rightarrow$ [ ] $\rightarrow 4$

In the next questions there are two machines. Work out the missing numbers in them.

5 $5 \rightarrow$ [ ] $\rightarrow 8 \rightarrow$ [ ] $\rightarrow a$
   $1 \rightarrow$ [$+3$] $\rightarrow b \rightarrow$ [$\times 2$] $\rightarrow c$
   $e \rightarrow$ [ ] $\rightarrow d \rightarrow$ [ ] $\rightarrow 8$

6 $0 \rightarrow$ [ ] $\rightarrow a \rightarrow$ [ ] $\rightarrow b$
   $5 \rightarrow$ [$\times 3$] $\rightarrow c \rightarrow$ [-4] $\rightarrow d$
   $f \rightarrow$ [ ] $\rightarrow e \rightarrow$ [ ] $\rightarrow 8$

In these next sets of machines you may also need to work out what function is in the machine.

7 $12 \rightarrow$ [ ] $\rightarrow a \rightarrow$ [ ] $\rightarrow 14$
   $8 \rightarrow$ [-5] $\rightarrow b \rightarrow$ [A] $\rightarrow 6$
   $c \rightarrow$ [ ] $\rightarrow 6 \rightarrow$ [ ] $\rightarrow d$

8
$$13 \rightarrow \boxed{} \rightarrow 16 \rightarrow$$

$$5 \rightarrow \boxed{E} \rightarrow 8 \rightarrow \boxed{F} \rightarrow 2$$

$$b \rightarrow \rightarrow 32 \rightarrow \rightarrow 8$$

⇨ Using symbols again

In the previous exercise the symbols stood for numbers. Sometimes you have puzzles where the symbols stand for operations.

Examples:

On the planet Antaeon the symbol ‡ stands for 'add them together and subtract 1'

So $3 ‡ 2 = 3 + 2 - 1$

$\qquad = 4$

(i) What is the value of: (a) $4 ‡ 3$

$\qquad$ (b) $5 ‡ (5 ‡ 2)$

$\qquad$ (c) $(3 ‡ 7) ‡ (5 ‡ 2)$?

(ii) If $a ‡ b = 5$ what are the possible values of a and b?

(i) (a) $4 ‡ 3 = 4 + 3 - 1$

$\qquad = 6$

(b) $5 ‡ (5 ‡ 2) = 5 ‡ (5 + 2 - 1)$

$\qquad = 5 ‡ 6$

$\qquad = 5 + 6 - 1$

$\qquad = 10$

(c) $(3 ‡ 7) ‡ (5 ‡ 2) = (3 + 7 - 1) ‡ (5 + 2 - 1)$

$\qquad = 9 ‡ 6$

$\qquad = 9 + 6 - 1$

$\qquad = 14$

(ii) If $a \ddagger b = 5$

$a + b - 1 = 5$

$a + b = 6$

Possible values of a and b are 1 and 5, 2 and 4, 3 and 3, 4 and 2, 5 and 1

Exercise 5.5

1 The symbol ♦ means 'add them together and multiply the answer by 3', so 5 ♦ 4 is $3 \times (5 + 4) = 27$

(a) Find the value of:

(i) 3 ♦ 1 (ii) 2 ♦ 2 (iii) 6 ♦ (5 ♦ 4) (iv) (4 ♦ 2) ♦ (5 ♦ 1)

(b) If $a ♦ b = 9$ and a and b are not equal what are the possible values of a and b?

2 The symbol △ means 'half of the sum' so 5 △ 3 means $(5 + 3) \div 2 = 4$

(a) Find the value of:

(i) 4 △ 2 (ii) 7 △ 2 (iii) 5 △ (4 △ 6) (iv) (10 △ 4) △ (12 △ 6)

(b) If $a △ b = 1\frac{1}{2}$ and a and b are not equal, what are the possible values of a and b?

3 The symbol ♣ means 'the product less the sum' so 5 ♣ 4 means $(5 \times 4) - (5 + 4) = 11$

(a) Find the value of:

(i) 3 ♣ 4 (ii) 5 ♣ 3 (iii) 5 ♣ (4 ♣ 2) (iv) 3 ♣ (3 ♣ (3 ♣ 3))

(b) If $a ♣ b = 0$ what are the possible values of a and b?

4 The symbol ♦ means 'double the sum then subtract 3'

(a) Find the value of:

(i) 4 ♦ 1 (ii) 7 ♦ 3 (iii) (2 ♦ 7) ♦ 5 (iv) (5 ♦ 1) ♦ (3 ♦ 3)

(b) If $a ♦ b = 5$ and a and b are not equal, what are the possible values of a and b?

5 (a) Look at these calculations and work out the meaning of the symbol ♦

$5 ♦ 2 = 20$ $6 ♦ 3 = 36$ $4 ♦ 7 = 56$

(b) Find the value of $4 ♦ 3$

6 (a) Look at these calculations and work out the meaning of the symbol ⊖

$5 ⊖ 2 = 20$ $6 ⊖ 3 = 26$ $4 ⊖ 7 = 32$

(b) Find the value of $4 ⊖ 3$

7 (a) Look at these calculations and work out the meaning of the symbol △

$4 △ 2 = 6$ $6 △ 3 = 15$ $8 △ 5 = 37$

(b) Find the value of $5 △ 3$

⇨ Sequences

A sequence is a list of numbers or objects that are in a special order. Each number is a **term**. If the terms are numbers, you can calculate the 'difference', that is how much larger or smaller one term is than the one before it.

Example:

Find the difference between the consecutive terms and then work out the next two terms in this sequence.

3, 5, 7, 9, 11, ...

When finding missing numbers in a sequence, you can write the difference between the terms with little arcs, like this.

```
 +2   +2   +2   +2    +2    +2
⌒    ⌒    ⌒    ⌒    ⌒     ⌒
3,   5,   7,   9,   11,   13,   15
```

The difference is $+ 2$ and the next two terms are 13 and 15

Some sequences are very simple, as in the example above, but in the next exercise there are some special sequences that you should look out for.

1 **Simple sequences**

Look at this sequence.

4, 7, 10, 13, 16, ...

(a) Write down the pattern of differences in this simple sequence.

(b) How would you describe this sequence?

(c) What are the next three numbers in the sequence?

2 **Square numbers**

(a) Count the dots in these patterns and then work out the next three numbers in the sequence.

(b) Write down the pattern of differences in the sequence of square numbers.

(c) How would you describe this sequence?

3 **Triangular numbers**

(a) Count the dots in these patterns and then work out the next three numbers in the sequence.

(b) Write down the pattern of differences in the sequence of triangular numbers.

(c) How would you describe this sequence?

4 **Fibonacci numbers**

The Fibonacci sequence is:

1, 1, 2, 3, 5, 8, 13, 21, 34, ...

This is a very special sequence. To find out why, you need to look at the differences.

(a) Write down the pattern of differences in the Fibonacci sequence.

(b) How would you describe this sequence?

(c) How would you work out the next number in the sequence?

(d) What is the next number?

5 Two-step sequences

Sometimes the pattern of differences in a sequence is not simple. You will need to use + and − to describe it.

Look at this sequence.

     5, 8, 7, 10, 9, 12, 11, ...

(a) Write down the pattern of differences in this sequence.

(b) How can you describe this sequence?

(c) What are the next three numbers in the sequence?

6 Embedded sequences

If none of the above seems to work, you may have two sequences embedded in one.

Look at this sequence.

     5, 4, 10, 7, 15, 10, 20, 13, ...

Look at it again, like this.

     5, 4, 10, 7, 15, 10, 20, 13, ...

(a) What is the pattern of differences in the red numbers?

(b) What is the pattern of differences in the black numbers?

(c) What are the next three numbers in the sequence?

⇨ Finding missing terms

In the next exercise you are going to find the missing terms in the sequences. Remember to work out the differences first and then see what type of sequence you have.

Examples:

Find the missing terms in each sequence.

  (i) 4, 7, 10, 13, ..., ...

 (ii) 0, 3, 8, 15, ..., ...

(iii) 2, 5, 7, 12, 19, ..., ...

(iv) 1, 3, 2, 6, 4, 9, ..., ...

        (i)
           +3 +3 +3 +3 +3
           4, 7, 10, 13, 16, 19

This is a simple sequence with a difference of + 3

       (ii)
          +3 +5 +7 +9 +11
         0, 3, 8, 15, 24, 35

The differences form a sequence of odd numbers so this is based on the sequence of square numbers.

These are 'square numbers − 1'

      (iii)
          +3 +2 +5 +7 +12 +19
         2, 5, 7, 12, 19, 31, 50

Aside from the first the differences are the same as the terms.

This is a Fibonacci-type sequence.

      (iv)
         +2 −1 +4 −2 +5
        1, 3, 2, 6, 4, 9, ..., ...

There is no obvious pattern to the differences.

Try alternate terms.

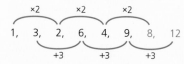

This is a + 3 sequence embedded in a × 2 sequence.

Exercise 5.7

Find the missing terms in each sequence.

1 6, 11, 16, 21, 26, ..., ...

2 4, 8, 12, 16, 20, ..., ...

3 4, 9, 16, 25, 36, ..., ...

4 2, 5, 10, 17, 26, ..., ...

5 1, 3, 6, 10, 15, ..., ...

6 0, 2, 5, 9, 14, ..., ...

7 1, 1, 2, 3, 5, 8, ..., ...

8 0, 3, 3, 6, 9, 15, ..., ...

9 7, 1, 9, 3, 11, 5, ..., ...

10 6, 1, 9, 3, 12, 9, ..., ...

11 $9\frac{1}{2}$, 8, $6\frac{1}{2}$, 5, $3\frac{1}{2}$, ..., ...

12 65, 59, 68, 62, 71, ..., ...

13 4, 2, 8, 4, 16, ..., ...

14 1, 4, 5, 9, 14, 23, ..., ...

More missing terms

In the next exercise the missing terms are not just at the end. As before, calculate the differences and spot the pattern. As you will be adding and subtracting, multiplying and dividing, take care to check your final answer.

Examples:

Find the missing terms in these sequences.

(i) 1, ..., 7, 10, ..., 16

(ii) ..., ..., 5, 8, 13, 21

(i)
```
     +3   +3   +3   +3   +3
    ⌒    ⌒    ⌒    ⌒    ⌒
  1,   4,   7,   10,  13,  16
```

This is a simple sequence with a difference of + 3

(ii)
```
     +1   +2   +3   +5   +8
    ⌒    ⌒    ⌒    ⌒    ⌒
  2,   3,   5,   8,   13,  21
```

The differences are the same as the terms, so this is a Fibonacci-type sequence.

Find the missing terms in each sequence.

1 1, 4, ..., ..., 25, 36, 49

2 ..., ..., 6, 10, 15, 21, 28

3 121, 100, ..., 64, ..., 36

4 2, ..., 12, 17, ..., 27

5 1, 3, ..., 7, ..., 18, 29

6 $13\frac{1}{2}$, 11, ..., 6, ..., 1

7 ..., 11, 3, 21, 5, ..., 7

8 81, 64, ..., 36, ..., 16

9 ..., 16, 3, 26, ..., 36, 5

10 ..., 19, $14\frac{1}{2}$, 10, ..., 1

11 ..., $2\frac{1}{2}$, 5, $8\frac{1}{2}$, ..., $18\frac{1}{2}$

12 2, 6, ..., 54, ..., 486

13 2, ..., 5, 8, 12, ..., 23

14 10, 7, 17, ..., 41, 65, ...

⇨ Number puzzles

For this last exercise, you need to use the information that you know about numbers to solve the puzzles.

1 I have two numbers with a sum of 11 and a product of 24
   What are the numbers?

2 I have two numbers with a sum of 17 and a product of 60
   What are the numbers?

3 I have two numbers with a difference of 4 and a product of 60
   What are the numbers?

4 I have two numbers with a difference of 13 and a product of 90
   What are the numbers?

5 (a) A number that is less than 20 has remainder 2 when divided
       by 3, and remainder 3 when divided by 4
       What is the number?

   (b) Write a calculation that includes 3 and 4 to explain your answer.

6 **(a)** A number that is less than 20 has remainder 2 when divided by 3, and remainder 4 when divided by 5
What is the number?

(b) Write a calculation that includes 3 and 5 to explain your answer.

7 **(a)** A number that is less than 100 has remainder 3 when divided by 4, remainder 4 when divided by 5 and remainder 5 when divided by 6
What is the number?

(b) Explain your answer.

8 When I divide a two-digit number by 4 the remainder is 2 and when I divide it by 7 the remainder is 5
What is the smallest number I could have started with?

9 When I divide a two-digit number by 6 the remainder is 4 and when I divide it by 8 the remainder is 6
What is the largest number I could have started with?

10 I remember the four digits on my padlock like this.

The last digit is four times the first, and the third digit is three times the second but one less than the fourth.

What is my number?

Activity – Consecutive number investigation

Consecutive numbers follow directly after one another, such as 5, 6, 7

$5 + 6 + 7 = 18$

Are there any other ways of adding consecutive numbers to make 18?

Repeat this exercise to find consecutive number sums for numbers 10 to 30

Write down any rules or patterns you notice.

6 More about numbers

In this chapter, you are going to revise some of the things that you already know about numbers. You will also use them in some new ways.

⇨ Index numbers

You know that if you measure shapes and solids in centimetres, then the units for area and volume are square centimetres (cm^2) and cubic centimetres (cm^3).

The small, raised 2 and the small, raised 3 **indicate** how many times centimetres have been multiplied by centimetres and these little numbers are called **index numbers**.

The same is true with numbers.

You can write:

● 4×4 as 4^2 and call it 'four squared'

● $2 \times 2 \times 2$ as 2^3 and call it 'two cubed'.

There is no word for index numbers above three, so you use the expression 'to the power...'. You would write:

● $5 \times 5 \times 5 \times 5 = 5^4$ and say 'five to the power 4'

$$5^4 = 625$$

● $10 \times 10 \times 10 \times 10 \times 10 = 10^5$ and say 'ten to the power 5'

$$10^5 = 100\ 000$$

You can see that as the powers get larger you will be dealing with some very large numbers. Therefore, it can be easier to write them as powers, for example, 10^5 rather than 100 000

Examples:

(i) What is three to the power five?

(ii) Write $5 \times 5 \times 5$ as a power of 5

(i) $3^5 = 3 \times 3 \times 3 \times 3 \times 3$

$\quad = 243$

(ii) $5 \times 5 \times 5 = 5^3$

Exercise 6.1

1 Work out the value of each number. Write it out as a power, with index numbers, first. Then calculate the value, as in the example above.

(a) six squared   (c) three to the power four

(b) five cubed   (d) two to the power five

2 Write each of these as a single digit with an index or power.

(a) $2 \times 2 \times 2$ (d) $7 \times 7 \times 7 \times 7 \times 7$

(b) 4×4 (e) $6 \times 6 \times 6 \times 6$

(c) $3 \times 3 \times 3$ (f) $4 \times 4 \times 4 \times 4 \times 4$

3 Work out the value of each number.

(a) 3^2 (c) 2^6

(b) 6^3 (d) 4^2

4 Copy each pair of numbers and write < or > between them.

(a) $3^2 \;\boxed{}\; 2^3$ (c) $5^2 \;\boxed{}\; 2^5$

(b) $5^3 \;\boxed{}\; 3^5$ (d) $3^3 \;\boxed{}\; 5^2$

5 Copy each pair of numbers and write <, > or = between them.

(a) $23 \boxed{} 2^3$

(c) $27 \boxed{} 2^7$

(b) $5^3 \boxed{} 53$

(d) $4^2 \boxed{} 2^4$

6 If $a^b = 1000$ what are the values of a and b?

7 If $a^b = 27$ what are the values of a and b?

8 If $a^b = 32$ what are the values of a and b?

9 What is the value of $3^2 + 4^2$? Write your answer as a square number.

10 What is the answer to $5^2 + 12^2$? Write your answer as a square number?

⇨ Square roots and cube roots

If you know that a number $a^2 = 9$ then you also know that:

$$a \times a = 9$$

and therefore $\qquad a = 3$

Then a is the **square root** of 9 and you write this as $\sqrt{9} = 3$

Other powers can also have roots.

If you know that a number $b^3 = 8$ then you also know that:

$$b \times b \times b = 8$$

and therefore $\qquad b = 2$

Then b is the **cube root** of 8 and you write this as $\sqrt[3]{8} = 2$

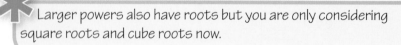

Larger powers also have roots but you are only considering square roots and cube roots now.

Exercise 6.2

Find the value of each square root or cube root.

1 $\sqrt{9}$

2 $\sqrt{100}$

3 $\sqrt[3]{27}$

4 $\sqrt{4}$

5 $\sqrt[3]{1000}$

6 $\sqrt[3]{125}$

7 $\sqrt{64}$

8 $\sqrt[3]{64}$

9 $\sqrt{81}$

10 $\sqrt[3]{216}$

11 $\sqrt{16}$

12 $\sqrt[3]{729}$

13 $\sqrt{49}$

14 $\sqrt{25}$

15 $\sqrt[3]{343}$

16 $\sqrt[3]{512}$

17 $\sqrt{36}$

18 $\sqrt{10\,000}$

19 $\sqrt{1\,000\,000}$

20 $\sqrt[3]{1\,000\,000}$

⇨ Factors and multiples

A number that divides exactly into another number is called a **factor** of that number. For example, **3** and **6** are **factors** of **12**

Numbers that are the result of a number (or factor) being multiplied by another whole number are called **multiples**. For example, **12** is a **multiple** of **6** – and also a multiple of 4 and of 3 and of 2

Whether you knew it or not, when you learnt your times tables you were studying factors and multiples.

● Multiples are answers to multiplications.

● Factors are answers to divisions with no remainder.

As **7 × 8 = 56** then:

● **56** is a **multiple** of **7** and of **8**

● **7** and **8** are **factors** of **56**

Finding factors by dividing

To find out if a number is a factor of another number, you need to see if it divides into it exactly without leaving a remainder.

Remember the rules of divisibility.

● All even numbers can be divided by 2

● If the digits of a number add up to a multiple of 3, the number can be divided by 3

● If the last two digits of a number can be divided by 4, the whole number can be divided by 4

● If the last digit of a number is 5 or 0, the whole number can be divided by 5

● If the digits of an even number add up to a multiple of 3, the number can be divided by 6

● If the last three digits of a number can be divided by 8, the whole number can be divided by 8

● If the digits of a number add up to a multiple of 9, the whole number can be divided by 9

● If the last digit of a number is 0, the whole number can be divided by 10

Example:

Are these numbers multiples of 3, 4, 6 or 9?

(i)   918   (ii) 7512

(i)   The digit sum of 918 is $9 + 1 + 8 = 18$

      18 is a multiple of 3 and 9

      918 is even but 18 cannot be divided by 4

      918 is a multiple of 3, 6, and 9 but not 4

(ii) The digit sum of 7512 is $7 + 5 + 1 + 2 = 15$

      15 is a multiple of 3 but not 9

      7512 is even and 12 can be divided by 4

      7512 is a multiple of 3, 4, and 6 but not 9

Exercise 6.3

1 Are these numbers multiples of 3, 4, 6 or 9?

   (a) 72   (b) 45   (c) 103   (d) 415   (e) 4107

2 Are these numbers multiples of 2, 6 or 8?

   (a) 36   (b) 86   (c) 141   (d) 294   (e) 434

3 Can 3456 be divided by:

   (a) 2   (b) 3   (c) 4   (d) 5   (e) 6?

4 Is 7 a factor of:

   (a) 98   (b) 133   (c) 317   (d) 7056?

5 What is the highest number that is a factor of both 12 and 15?

6 What is the lowest number that has both 3 and 7 as factors?

7 What is the lowest number that can be divided by both 4 and 6?

8 Which number between 100 and 200 is divisible by both 8 and 9?

9 Which numbers between 100 and 200 are divisible by both 6 and 9?

10 A three-digit number is a multiple of 3 and of 7 and the sum of its digits is 12
   What is the number?

11 (a) What is the lowest number that is a multiple of 2, 3, 4, 5, 6, 8 and 9?

   (b) Write down all its other factors.

12 Now you know more about factors and divisibility, write a rule of divisibility for multiples of 12

⇨ Prime numbers

Consider the factors of the numbers 7, 11, 23 and 71

- 7 has only two factors: 1 and 7

- 11 has only two factors: 1 and 11

- 23 has only two factors: 1 and 23

- 71 has only two factors: 1 and 71

Numbers that have only two factors, the number itself and 1, are prime numbers.

Exercise 6.4

1 Write down all the prime numbers between 1 and 50

2 Write down all the prime numbers between 51 and 100

3 Copy this number grid, on squared paper.

101	102	103	104	105	106	107	108	109	110
111	112	113	114	115	116	117	118	119	120
121	122	123	124	125	126	127	128	129	130
131	132	133	134	135	136	137	138	139	140
141	142	143	144	145	146	147	148	149	150
151	152	153	154	155	156	157	158	159	160
161	162	163	164	165	166	167	168	169	170
171	172	173	174	175	176	177	178	179	180
181	182	183	184	185	186	187	188	189	190
191	192	193	194	195	196	197	198	199	200

Use your number square to find all the prime numbers, from 101 to 200

(a) Cross out all the multiples of 2 and all the multiples of 5

(b) Cross out all the multiples of 3

(c) Cross out all the multiples of 7

(d) Cross out all the multiples of 11 and all the multiples of 13

If you have done this correctly, you will just be left with prime numbers.

⇨ Prime factors

The prime factors of a number are its **factors** that are also **prime numbers**.

If a number is not prime, it is a composite number. Composite numbers can be broken down, or written as the product of two or more other numbers.

It follows that every number that is not prime can be written as a product of prime numbers.

You can find the prime factors of a number in two ways, as described below.

Factor trees

A factor tree is a way of finding prime factors by repeatedly breaking the number and then its factors down.

Every time you find a factor that is not a prime number, you break it down again.

Example:

Write 36 as the product of its prime factors.

Here are three different ways of arriving at the answer.

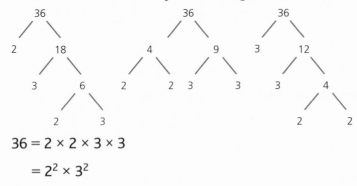

$36 = 2 \times 2 \times 3 \times 3$

$\quad = 2^2 \times 3^2$

Ladder division

Ladder division involves dividing by a succession of prime numbers until the answer is a prime number.

If it is an even number, start by dividing by 2 and then continue dividing by 2 until the answer is odd. Then find the next smallest prime factor, divide by that until you cannot divide by it any more. Continue with the next smallest prime factor, and so on.

Example:

Find the prime factors of 480

2	4 8 0
2	2 4 0
2	1 2 0
2	6 0
2	3 0
3	1 5
5	5
	1

Factor trees work well for smaller numbers with fewer factors. Ladder division is a better method for larger numbers.

$480 = 2 \times 2 \times 2 \times 2 \times 2 \times 3 \times 5$

$\quad = 2^5 \times 3 \times 5$

Note that the final answer is a product, with repeated prime factors written as index numbers. That helps you to double check that you have calculated and written down the correct number of factors.

Exercise 6.5

1 Use a factor tree to express each of the numbers below as the product of its prime factors. Write your answers in index form.

(a) 18   (c) 84   (e) 150

(b) 56   (d) 72   (f) 100

2 Use ladder division to express each of the numbers below as the product of its prime factors. Write your answers in index form.

(a) 240   (c) 180   (e) 275

(b) 450   (d) 135   (f) 208

3 Use either method to express each of the numbers below as the product of its prime factors. Write your answers in index form.

(a) 155   (c) 260   (e) 215

(b) 143   (d) 330   (f) 323

⇨ Highest common factors

Think about the numbers 100 and 150

$$100 = ②\times 2 \times ⑤ \times ⑤$$
$$150 = ②\times 3 \times ⑤ \times ⑤$$

You can see that they have some prime factors in common.

$$2 \times 5 \times 5 = 50$$

This means that the highest number that divides into both 100 and 150 is 50

$$100 \div 50 = 2$$

$$150 \div 50 = 3$$

So 50 is the highest common factor of 100 and 150

A Venn diagram is a useful way to find the highest common factor of two or more numbers. Each loop in the Venn diagram shows the factors of one of the numbers.

The regions where the loops overlap show the common factors.

Examples:

(i) Use a Venn diagram to find the highest common factor of 18 and 48

(ii) Use a Venn diagram to find the highest common factor of 84, 260 and 330

(i) $18 = 2 \times 3 \times 3$

$\quad = 2 \times 3^2$

$48 = 2 \times 2 \times 2 \times 2 \times 3$

$\quad = 2^4 \times 3$

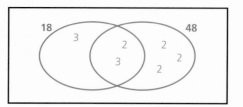

2 and 3 are common factors of 18 and 48

The highest common factor is $2 \times 3 = 6$

(ii) $84 = 2 \times 2 \times 3 \times 7$

$\quad = 2^2 \times 3 \times 7$

$260 = 2 \times 2 \times 5 \times 13$

$\quad = 2^2 \times 5 \times 13$

$330 = 2 \times 3 \times 5 \times 11$

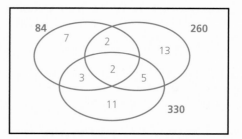

2 is the highest common factor of 84, 260 and 330

> *The **highest common factor** of two or more numbers is the biggest number that can be divided exactly into all of them.*

Exercise 6.6

Use a Venn diagram to find the highest common factor of each group of numbers.

If you have worked out the number as a product of prime factors in an earlier exercise, you do not need to show your working again.

1 (a) 18 and 24   (d) 12 and 42

   (b) 72 and 84   (e) 88 and 132

   (c) 84 and 100   (f) 120 and 135

2 **(a)** 240 and 450

   (b) 180 and 450

   (c) 150 and 180

   (d) 135 and 275

   (e) 208 and 275

   (f) 130 and 275

3 **(a)** 12, 15 and 18

   (b) 24, 36 and 54

   (c) 72, 180 and 240

   (d) 25, 30 and 85

   (e) 30, 75 and 105

   (f) 240, 320 and 345

4 What is the largest number that will divide exactly into 36, 90 and 150?

5 What is the largest number that will divide exactly into 75, 105 and 180?

6 What is the largest number that will divide exactly into 144, 180 and 288?

7 I have two lengths of fabric, one is 24 cm wide and one is 18 cm wide. I want to cut the fabric up into ribbons of equal width. What is the greatest possible width of my ribbons?

8 The deputy head has decided to reorganise the school day. There will now be 2 hours 55 minutes of lessons before lunch and 1 hour 45 minutes of lessons after lunch.

   (a) If all lessons are the same length what is the longest possible length of a lesson?

   (b) How many lessons will there be?

⇨ Lowest common multiples

Look at these multiples of 7 and 8

   7, 14, 21, 28, 35, 42, 49, (56), 63, ...

   8, 16, 24, 32, 40, 48, (56), 64, 72, ...

You can see that 56 is the first number that occurs in both lists. It is the lowest common multiple of 7 and 8

   $56 = 7 \times 8$

You can always find a common multiple of two numbers by multiplying them together, but is that always the lowest?

Look again at the Venn diagram showing the factors of 18 and 48

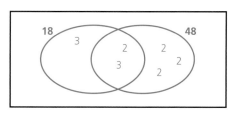

The **lowest common multiple** of two or more numbers is the lowest number that can be divided exactly by all of them.

$18 = \mathbf{2} \times \mathbf{3} \times 3$ $48 = 2 \times 2 \times 2 \times \mathbf{2} \times \mathbf{3}$

A common multiple of 18 and 48 must include all of their factors, but because 2×3 is a common factor you only need to include it once.

The lowest common multiple of 18 and 48 is
$2 \times 2 \times 2 \times 2 \times 3 \times 3 = 144$

Exercise 6.7

Use a Venn diagram the find the lowest common multiple of each group of numbers.

If you have worked out the number as a product of prime factors in an earlier exercise, you do not need to show your working again.

1 (a) 4 and 6
  (b) 8 and 12
  (c) 10 and 15

  (d) 15 and 25
  (e) 12 and 18
  (f) 16 and 24

2 (a) 24 and 36
  (b) 24 and 56
  (c) 30 and 45

  (d) 45 and 65
  (e) 208 and 275
  (f) 130 and 275

3 (a) 10, 15 and 24
  (b) 24, 36 and 60
  (c) 72, 120 and 180

  (d) 75, 100 and 125
  (e) 60, 90 and 200
  (f) 240, 360 and 400

4 What is the smallest number that can be divided exactly by 6, 9 and 15?

5 What is the smallest number that can be divided exactly by 36, 75 and 90?

6 What is the smallest number that can be divided exactly by 39, 52 and 130?

7 A pair of bells is rung together at 9:00 a.m. Bell A is then rung every 3 minutes and Bell B every 5 minutes.

   (a) What will be the next time that they ring together?

   (b) How many times will they ring together from 9am to 12 noon?

8 Ben and Ali start a long-distance running race at the same time. Ben completes a lap every 5 minutes and Ali every 6 minutes. When will they next meet at the starting point?

9 I have a pile of 20p coins and a pile of 50p coins. If my piles both have the same value, what is the least amount of money I could have in total?

10 The lowest common multiple of three numbers is 156 and their sum is 117. What are the three numbers?

Exercise 6.8: Summary exercise

1 Write down the value of each number.

   (a) 5^2

   (c) $\sqrt[3]{27}$

   (b) $\sqrt{144}$

   (d) 3^4

2 Write down the value of each expression.

   (a) $2^3 \times 3$

   (c) $2 \times 3^2 \times 7$

   (b) 3×5^2

   (d) $2^3 \times 3^2$

3 Find the prime factors and hence the highest common factor of each set of numbers.

   (a) 8 and 10

   (c) 100 and 115

   (b) 12 and 16

   (d) 75, 120 and 140

4 Find the prime factors and hence the lowest common multiple of each set of numbers.

(a) 4 and 9

(c) 35 and 42

(b) 18 and 24

(d) 12, 15 and 20

5 Two racing cars start at the same time. Car A completes a lap in 15 seconds and Car B completes a lap in 18 seconds.

(a) After how many seconds will they next be at the starting line at the same time?

(b) How many more laps will Car A have done by then?

6 Two pieces of wire, one 312 cm long and one 468 cm long, are cut into equal lengths, with no wire left over. What is the longest possible length of the pieces of wire?

7 I have 36 red dots and 45 blue dots and I want to make a row of identical patterns with a mix of red and blue dots. If I use all my dots how many patterns can I make and how many red and blue dots are in each pattern?

8 On a special clock a little bird comes out of his house every 15 minutes and a little cat every 25 minutes. If they come out together at 12 noon, what is the next time both the bird and the cat appear?

Activity – 100 prisoners

100 prisoners are locked in 100 cells. The cells are numbered 1 to 100

On day 1, the guard turns the keys in all the cells so that they are all unlocked.

On day 2, the guard turns the key in every cell with a number that is a multiple of 2. This locks all the even-numbered cells.

On day 3, the guard turns the key in every cell with a number that is a multiple of 3, locking or unlocking them.

On day 4, the guard turns the key in every cell with a number that is a multiple of 4, locking or unlocking them.

This goes on for 100 days and on the 100th day every prisoner in an unlocked cell can go free.

Which prisoners can go free?

Copy this 1 to 100 square on to squared paper.

Use your copy to help you with this problem.

1	2	3	4	5	6	7	8	9	10
11	12	13	14	15	16	17	18	19	20
21	22	23	24	25	26	27	28	29	30
31	32	33	34	35	36	37	38	39	40
41	42	43	44	45	46	47	48	49	50
51	52	53	54	55	56	57	58	59	60
61	62	63	64	65	66	67	68	69	70
71	72	73	74	75	76	77	78	79	80
81	82	83	84	85	86	87	88	89	90
91	92	93	94	95	96	97	98	99	100

Getting started

Have a look at cells 1 to 10

On day 1, all cells are unlocked so leave them unmarked.

On day 2, cells 2, 4, 6, 8 and 10 are locked, so put a diagonal line through them.

On day 3 the locks are turned in cells 3, 6 and 9 so 3 and 9 are now locked. Put a diagonal line through them. Cell 6 is unlocked, so put a line the other way.

| 1 | 2 | 3 | 4 | 5 | 6 | 7 | 8 | 9 | 10 |

Continue for 10 days.

1 How many cells have been locked or unlocked an odd number of times? Are they now open or still locked?

2 How many cells have been locked or unlocked an even number of times? Are they now open or still locked?

Continue for 100 days.

Take care to record carefully whether the cells have been locked or unlocked an even number or an odd number of times.

3 After 100 days, how many prisoners are set free?

4 What are the numbers of the cells of the freed prisoners?

5 Are these a special type of number?

6 Explain how it is these prisoners are freed. You may find words such as 'odd', 'even', 'prime', 'factor' and 'multiple' useful!

7 Some cells have only had the door locked once and never had it unlocked again. What numbers are these cells?

8 Are these a special type of number?

9 Explain why these cells were never unlocked again. You may find words such as 'odd', 'even', 'prime', 'factor' and 'multiple' useful again.

7 Fractions

1/7/2020

You already use fractions all of the time.

● The time is half past three.

● Ben is ten and a half years old.

● The shop is quarter of a mile from here.

In this chapter, you will review and practise what you know, and then go on to compare and calculate with fractions.

⇨ What do you know?

Numerator and denominator

The bottom number of a fraction is called the denominator. It is very important because it tells you the number of equal parts there are in the whole.

The top number is the numerator. It tells you how many of these equal parts you have.

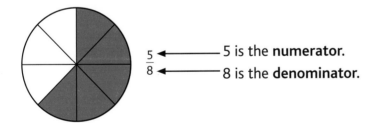

$\frac{5}{8}$ 5 is the **numerator**.
       8 is the **denominator**.

A **proper fraction** is one in which the numerator is smaller than the denominator. It has a value between zero and one. For example, $\frac{2}{5}$ is a proper fraction.

Lowest terms

Sometimes you will notice that a fraction you are looking at can be written more simply.

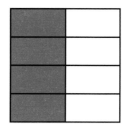

$$\frac{4}{8} = \frac{1}{2}$$

Although $\frac{4}{8}$ of the shape have been shaded, you can see that this equals one half, $\frac{1}{2}$

You **simplify** a fraction by dividing the numerator and denominator by a **common factor**.

Example:

Simplify the fraction $\frac{4}{8}$

$$\frac{4}{8} = \frac{4 \div 2}{8 \div 2}$$

$$= \frac{2 \div 2}{4 \div 2}$$

$$= \frac{1}{2}$$

When you cannot simplify a fraction any more, it is in its **lowest terms** (simplest form).

Equivalent fractions

Look at these two fractions.

$$\frac{5}{6} \qquad\qquad \frac{15}{18}$$

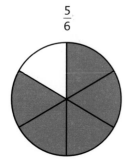 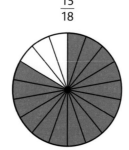

From the diagram, you can see that the fractions are both the same size, even though they have different denominators. Fractions that are the same size are equivalent fractions.

Given any fraction, you can find fractions that are equivalent to it by multiplying or dividing the numerator and denominator by the same number.

Example:

Find the missing number, if $\frac{5}{6} = \frac{\square}{18}$

Compare the denominators.

$6 \times 3 = 18$, so to get from $\frac{5}{6}$ to $\frac{\square}{18}$ you multiply the numerator and denominator by 3

$$\frac{5 \times 3}{6 \times 3} = \frac{15}{18}$$

You can use equivalent fractions to compare fractions.

Example:

Which is larger, $\frac{5}{6}$ or $\frac{4}{5}$? $= \overline{} \; 30$

You need to find equivalent fractions, with the same denominators, for both $\frac{5}{6}$ and $\frac{4}{5}$

To do this, look for the lowest number that is a multiple of both 5 and 6. This is their **lowest common multiple**. This will be the **lowest common denominator**.

The lowest common multiple of 5 and 6 is 30

For $\frac{5}{6}$: $\frac{5 \times 5}{6 \times 5} = \frac{25}{30}$

For $\frac{4}{5}$: $\frac{4 \times 6}{5 \times 6} = \frac{24}{30}$

Therefore $\frac{5}{6}$ is larger than $\frac{4}{5}$ so you can write $\frac{5}{6} > \frac{4}{5}$

Fractions greater than one

There are two ways of writing fractions that are greater than one.

- In an **improper fraction** the numerator is bigger than the denominator. For example, $\frac{12}{5}$ is an improper fraction.

- A **mixed number** is made up of a whole number and a **proper fraction**. For example, $2\frac{2}{5}$ is a mixed number.

 Improper fractions are sometimes called **top-heavy fractions**. Can you see why?

17/7/2020

Exercise 7.1

1 Simplify each of these fractions by dividing the numerator and denominator by a common factor. Write them in their lowest terms.

(a) $\frac{8}{24}$

(b) $\frac{3}{12}$

(c) $\frac{10}{15}$

(d) $\frac{18}{24}$

(e) $\frac{40}{50}$

(f) $\frac{40}{60}$

(g) $\frac{12}{18}$

(h) $\frac{24}{72}$

(i) $\frac{65}{100}$

(j) $\frac{45}{135}$

2 Copy the fractions and fill in the missing numerators to make the fractions equivalent.

(a) $\frac{4}{5} = \frac{\square}{25}$

(b) $\frac{7}{8} = \frac{\square}{16}$

(c) $\frac{3}{5} = \frac{\square}{20}$

(d) $\frac{6}{7} = \frac{\square}{21}$

(e) $\frac{3}{4} = \frac{\square}{24}$

(f) $\frac{7}{15} = \frac{\square}{30}$

3 Copy the fractions and fill in the missing denominators to make the fractions equivalent.

(a) $\dfrac{3}{4} = \dfrac{12}{\rule{1em}{0.4pt}}$

(d) $\dfrac{5}{\rule{1em}{0.4pt}} = \dfrac{20}{36}$

(b) $\dfrac{2}{3} = \dfrac{12}{\rule{1em}{0.4pt}}$

(e) $\dfrac{14}{\rule{1em}{0.4pt}} = \dfrac{7}{24}$

(c) $\dfrac{12}{\square} = \dfrac{3}{20}$

(f) $\dfrac{7}{45} = \dfrac{28}{\square}$

4 In each of these, multiply or divide the numerator and denominator by the same number to find the equivalent fraction.

(a) $\dfrac{3}{4} = \dfrac{\rule{1em}{0.4pt}}{36}$

(d) $\dfrac{7}{10} = \dfrac{\rule{1em}{0.4pt}}{40}$

(b) $\dfrac{5}{\rule{1em}{0.4pt}} = \dfrac{25}{100}$

(e) $\dfrac{4}{15} = \dfrac{16}{\rule{1em}{0.4pt}}$

(c) $\dfrac{\square}{25} = \dfrac{36}{100}$

(f) $\dfrac{3}{4} = \dfrac{\square}{360}$

5 There are 120 sweets in a jar. 25 of them are orange. What fraction of the sweets is this?

6 There are 750 people in the audience at a theatre. 360 of them are children. What fraction of the audience is made up of children?

7 There are 160 cars in a car park. 90 of them are blue. What fraction of the total number of cars are blue?

8 There are 24 pupils in the class. 15 of them are boys. What fraction of the class is boys?

9 32 children go on a school trip, 14 are boys. What fraction of the party are girls?

10 In a bag of 48 potatoes, 12 are rotten. What fraction of the potatoes is not rotten?

⇨ Ordering fractions

Examples:

(i) Write $\frac{9}{10}$, $\frac{4}{5}$ and $\frac{7}{8}$ in order, smallest first.

(ii) Write $1\frac{1}{3}$, $\frac{11}{9}$ and $1\frac{1}{5}$ in order, smallest first.

(i) The denominators of $\frac{9}{10}$, $\frac{4}{5}$ and $\frac{7}{8}$ are 5, 8 and 10

Comparing factors:

$$5 = 5 \times 1 \qquad 8 = 2 \times 2 \times 2 \qquad 10 = 2 \times 5$$

They will all divide into the number that is $2 \times 2 \times 2 \times 5 = 40$

For $\frac{9}{10} : \frac{9 \times 4}{10 \times 4} = \frac{36}{40}$

40 is the lowest common multiple of 5, 8 and 10

For $\frac{4}{5} : \frac{4 \times 8}{5 \times 8} = \frac{32}{40}$

For $\frac{7}{8} : \frac{7 \times 5}{8 \times 5} = \frac{35}{40}$

In order, the fractions are $\frac{4}{5}$, $\frac{7}{8}$, $\frac{9}{10}$

(ii) The denominators of $1\frac{1}{3}$, $\frac{11}{9}$ and $1\frac{1}{5}$ are 3, 5 and 9

These numbers can all be made from the factors 3, 3 and 5

They will all divide into the number that is $3 \times 3 \times 5 = 45$

For $1\frac{1}{3} : 1\frac{1 \times 15}{3 \times 15} = 1\frac{15}{45}$

45 is the lowest common multiple of 3, 9 and 15

For $\frac{11}{9} : 1\frac{2}{9} = 1\frac{2 \times 5}{9 \times 5}$ which is $1\frac{10}{45}$

For $1\frac{1}{5} : 1\frac{1 \times 9}{5 \times 9} = 1\frac{9}{45}$

In order, the fractions are $1\frac{1}{5}$, $\frac{11}{9}$, $1\frac{1}{3}$

Exercise 7.2

1 Copy each pair of fractions, and then write < or > between
  them. You will need to find the lowest common denominator
  and calculate equivalent fractions.

(a) $\frac{1}{4}$ ☐ $\frac{2}{7}$ (f) $\frac{19}{8}$ ☐ $2\frac{5}{18}$

(b) $\frac{3}{4}$ ☐ $\frac{2}{3}$ (g) $\frac{13}{10}$ ☐ $1\frac{7}{15}$

(c) $\frac{3}{7}$ ☐ $\frac{5}{14}$ (h) $\frac{25}{14}$ ☐ $1\frac{15}{21}$

(d) $\frac{7}{12}$ ☐ $\frac{5}{8}$ (i) $1\frac{7}{15}$ ☐ $\frac{36}{25}$

(e) $\frac{3}{4}$ ☐ $\frac{7}{10}$ (j) $\frac{61}{24}$ ☐ $2\frac{19}{36}$

2 Write the fractions in each set in order, smallest first.

(a) $\frac{3}{5}, \frac{7}{10}, \frac{13}{25}$ (d) $\frac{13}{20}, \frac{8}{15}, \frac{3}{5}$

(b) $\frac{5}{6}, \frac{2}{3}, \frac{7}{9}$ (e) $2\frac{3}{4}, 2\frac{11}{12}, \frac{8}{3}$

(c) $1\frac{5}{8}, \frac{7}{4}, 1\frac{9}{16}$ (f) $\frac{7}{4}, \frac{5}{3}, \frac{9}{5}$

3 Write the fractions in each set in order, largest first.

(a) $\frac{5}{9}, \frac{2}{3}, \frac{7}{12}$ (d) $\frac{15}{16}, \frac{7}{8}, \frac{11}{12}$

(b) $1\frac{3}{10}, \frac{7}{5}, 1\frac{4}{15}$ (e) $\frac{17}{12}, 1\frac{4}{9}, 1\frac{3}{8}$

(c) $\frac{4}{3}, \frac{11}{9}, \frac{5}{4}$ (f) $1\frac{7}{10}, \frac{22}{15}, \frac{11}{6}$

⇨ Adding fractions

Adding fractions with the same denominator is straightforward.

To add fractions with different denominators, first you must write them as equivalent fractions with a common denominator.

Examples:

(i) Add: $1\frac{1}{6} + 1\frac{7}{12}$

(ii) Add: $1\frac{3}{4} + 2\frac{4}{5}$

(i) $1\frac{1}{6} + 1\frac{7}{12}$

12 will be the lowest common denominator for sixths and twelfths.

$$\frac{1}{6} = \frac{2}{12}$$ Find equivalent fractions.

$$1\frac{1}{6} + 1\frac{7}{12} = 2\frac{2+7}{12}$$ Add the whole numbers first.

$$= 2\frac{9}{12}$$

$$= 2\frac{3}{4}$$ Write the answer in its lowest terms.

(ii) $1\frac{3}{4} + 2\frac{4}{5}$

20 will be the lowest common denominator for quarters and fifths.

$$\frac{3}{4} = \frac{15}{20} \text{ and } \frac{4}{5} = \frac{16}{20}$$ Find equivalent fractions.

$$1\frac{3}{4} + 2\frac{4}{5} = 3\frac{15+16}{20}$$ Add the whole numbers first.

$$= 3\frac{31}{20}$$ Add the fractions.

$$= 3 + 1 + \frac{11}{20}$$ Turn the improper fraction into a mixed number.

$$= 4\frac{11}{20}$$ Add the whole numbers together.

Exercise 7.3

Add these fractions. Make sure the fractions in your answers are in their lowest terms.

1 $\frac{1}{2} + \frac{1}{6}$

2 $1\frac{1}{3} + 2\frac{5}{12}$

3 $\frac{3}{4} + 1\frac{5}{12}$

4 $1\frac{4}{15} + 2\frac{2}{3}$

5 $\frac{2}{7} + \frac{10}{21}$

6 $\frac{3}{4} + \frac{7}{12}$

7 $2\frac{7}{15} + 1\frac{2}{3}$

8 $1\frac{3}{10} + 2\frac{1}{2}$

9 $\frac{3}{4} + \frac{1}{3} + \frac{5}{12}$

10 $\frac{3}{5} + \frac{7}{15} + \frac{2}{3}$

11 $\frac{2}{3} + \frac{3}{5}$

12 $\frac{3}{4} + \frac{1}{6}$

13 $1\frac{2}{5} + 2\frac{3}{4}$

14 $1\frac{5}{6} + \frac{4}{9}$

15 $2\frac{5}{12} + 3\frac{3}{16}$

16 $2\frac{3}{4} + \frac{9}{14}$

17 $2\frac{4}{15} + 1\frac{7}{10}$

18 $1\frac{3}{8} + 4\frac{11}{18}$

19 $2\frac{7}{8} + 3\frac{7}{12}$

20 $4\frac{5}{6} + 1\frac{7}{8}$

21 $1\frac{7}{10} + 1\frac{3}{4}$

22 $1\frac{1}{4} + \frac{2}{3} + 2\frac{5}{12}$

23 $2\frac{4}{15} + 3\frac{2}{3} + 1\frac{1}{5}$

24 $1\frac{5}{16} + \frac{3}{8} + 1\frac{1}{6}$

25 $\frac{5}{12} + \frac{3}{16} + \frac{13}{24}$

26 $\frac{7}{10} + \frac{11}{15} + \frac{3}{4}$

27 $2\frac{11}{12} + 1\frac{3}{8} + 1\frac{4}{9}$

28 $2\frac{8}{15} + 2\frac{5}{12} + 1\frac{11}{20}$

29 $1\frac{7}{8} + 2\frac{3}{10} + 1\frac{5}{6}$

30 $3\frac{3}{4} + 2\frac{4}{15} + 4\frac{1}{6}$

⇨ Subtracting fractions

Just as for addition, when you have to subtract fractions with different denominators, you first need to write them as equivalent fractions with a common denominator.

Example:

(i) Subtract: $3\frac{2}{3} - 1\frac{1}{6}$

(ii) Subtract: $4\frac{5}{6} - 1\frac{1}{4}$

(i) $3\frac{2}{3} - 1\frac{1}{6}$

6 is the lowest common denominator for thirds and sixths.

$\frac{2}{3} = \frac{4}{6}$ Find equivalent fractions.

$3\frac{2}{3} - 1\frac{1}{6} = 2\frac{4-1}{6}$ Subtract the whole numbers first.

$= 2\frac{3}{6}$

$= 2\frac{1}{2}$ Write the answer as a fraction in its lowest terms.

(ii) $4\frac{5}{6} - 1\frac{1}{4}$

12 is the smallest multiple of both 4 and 6

$\frac{5}{6} = \frac{10}{12}$ and $\frac{1}{4} = \frac{3}{12}$

$4\frac{5}{6} - 1\frac{1}{4} = 3\frac{10-3}{12}$ Subtract the whole numbers first.

$= 3\frac{7}{12}$

Exercise 7.4

Subtract these fractions. Make sure the fractions in your answers are in their lowest terms.

1 $1\frac{1}{3} - \frac{1}{6}$

2 $2\frac{1}{2} - \frac{1}{12}$

3 $3\frac{3}{4} - 1\frac{5}{12}$

4 $2\frac{11}{15} - \frac{2}{5}$

5 $1\frac{5}{8} - \frac{1}{4}$

6 $3\frac{7}{12} - 1\frac{1}{4}$

7 $2\frac{8}{15} - \frac{1}{3}$

8 $3\frac{7}{8} - 1\frac{3}{4}$

9 $\frac{11}{12} - \frac{5}{6}$

10 $3\frac{17}{18} - 1\frac{2}{3}$

11 $\frac{3}{4} - \frac{1}{3}$

12 $\frac{4}{5} - \frac{3}{4}$

13 $1\frac{2}{5} - \frac{1}{4}$

14 $3\frac{5}{6} - 1\frac{1}{4}$

15 $5\frac{3}{4} - 2\frac{7}{10}$

16 $1\frac{5}{6} - \frac{4}{9}$

17 $3\frac{7}{15} - 2\frac{1}{6}$

18 $4\frac{9}{10} - 1\frac{4}{15}$

19 $3\frac{8}{9} - 1\frac{5}{6}$

20 $2\frac{5}{8} - 1\frac{1}{6}$

What happens when the fraction in the second number is bigger than the fraction in the first number?

One method is to turn a mixed number into an improper fraction before you subtract.

Example:

Subtract: $1\frac{1}{2} - \frac{5}{6}$

6 is the lowest common denominator.

$1\frac{1}{2} - \frac{5}{6} = 1\frac{3}{6} - \frac{5}{6}$ Find equivalent fractions. You cannot take 5 from 3

$= \frac{9}{6} - \frac{5}{6}$ Turn the mixed number into an improper fraction.

$= \frac{4}{6}$

$= \frac{2}{3}$

Another method is to 'change' a whole number to a fraction.

Example:

Subtract: $4\frac{1}{6} - 2\frac{3}{4}$

12 is the lowest common denominator.

$\frac{1}{6} = \frac{2}{12}$ and $\frac{3}{4} = \frac{9}{12}$ Find equivalent fractions.

$4\frac{1}{6} - 2\frac{3}{4} = 2\frac{2-9}{12}$ Subtract the whole numbers.

$\quad\quad\quad = 1 + \frac{12}{12} + \frac{2-9}{12}$ Change a whole number into $\frac{12}{12}$

$\quad\quad\quad = 1\frac{14-9}{12}$ Add $\frac{12}{12}$ to $\frac{2}{12}$

$\quad\quad\quad = 1\frac{5}{12}$ Take 9 from 14

Exercise 7.5

Subtract these fractions. Make sure the fractions in your answers are in their lowest terms.

1   $1\frac{1}{2} - \frac{3}{4}$ 6 $2\frac{5}{12} - 1\frac{3}{4}$

2   $1\frac{1}{3} - \frac{5}{6}$ 7 $4\frac{1}{5} - 3\frac{7}{10}$

3   $1\frac{5}{12} - \frac{3}{4}$ 8 $2\frac{3}{10} - \frac{4}{5}$

4   $1\frac{2}{5} - \frac{7}{10}$ 9 $3\frac{5}{12} - 1\frac{5}{6}$

5   $1\frac{1}{8} - \frac{3}{4}$ 10 $4\frac{3}{5} - 1\frac{7}{10}$

11   $1\frac{1}{5} - \frac{2}{3}$ 16 $1\frac{1}{8} - \frac{11}{12}$

12   $2\frac{1}{3} - 1\frac{3}{4}$ 17 $3\frac{3}{10} - 1\frac{7}{12}$

13   $2\frac{3}{10} - 1\frac{3}{4}$ 18 $3\frac{7}{10} - \frac{11}{15}$

14   $3\frac{1}{6} - 1\frac{4}{9}$ 19 $1\frac{4}{9} - \frac{5}{6}$

15   $2\frac{1}{4} - 1\frac{5}{6}$ 20 $3\frac{3}{8} - 1\frac{7}{10}$

⇨ Fraction patterns

You already know how to recognise patterns in numbers, and work
out missing terms. Now try some sequences that include fractions.
The method is the same.

Example:

Find the missing terms in this sequence.

$$2\frac{1}{6}, 3\frac{1}{3}, 4\frac{1}{2}, 5\frac{2}{3}, ..., ...$$

Start by finding the difference between terms.

6 is the lowest common denominator.

$$2\frac{1}{6}, 3\frac{2}{6}, 4\frac{3}{6}, 5\frac{4}{6}, ..., ...$$

The difference between terms is $1\frac{1}{6}$

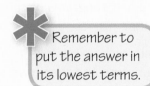

Remember to
put the answer in
its lowest terms.

$$5\frac{4}{6} + 1\frac{1}{6} = 6\frac{5}{6} \qquad 6\frac{5}{6} + 1\frac{1}{6} = 7\frac{6}{6} = 8$$

The missing terms are $6\frac{5}{6}$ and 8

Exercise 7.6

Find the missing terms in these patterns. You will need to find the
difference between terms first. Always write the fractions in their
lowest terms.

1 $\frac{1}{4}, \frac{1}{2}, \frac{3}{4}, 1, ..., ...$

2 $\frac{5}{6}, 1\frac{1}{6}, 1\frac{1}{2}, 1\frac{5}{6}, ..., ...$

3 $\frac{3}{10}, \frac{1}{2}, \frac{7}{10}, \frac{9}{10}, ..., ...$

4 $\frac{5}{12}, \frac{2}{3}, \frac{11}{12}, 1\frac{1}{6}, ..., ...$

5 $1\frac{5}{16}, 1\frac{3}{8}, 1\frac{7}{16}, 1\frac{1}{2}, ..., ...$

6 $3\frac{1}{2}, 3\frac{3}{4}, ..., 4\frac{1}{4}, 4\frac{1}{2}, ...$

7 $2\frac{1}{2}, 2\frac{5}{6}, ..., 3\frac{1}{2}, ..., 4\frac{1}{6}$

8 $4\frac{1}{4}, ..., 5\frac{1}{12}, 5\frac{1}{2}, ..., 6\frac{1}{3}$

9 $..., ..., 3\frac{5}{8}, 3\frac{13}{16}, 4, 4\frac{3}{16}$

10 $..., 5\frac{2}{5}, 5\frac{11}{20}, 5\frac{7}{10}, ..., 6$

Problem solving

Now you have learnt how to calculate with fractions here are some problems for you to solve.

Always write down all your working.

Exercise 7.7

Calculate the answers, showing all your working. If your answer is not a whole number write it as a fraction.

1 I put one and a quarter of a kilogram of potatoes in my bag and then add two and a half kilograms of carrots. What is the total mass of my shopping?

2 I walk for half a mile, pick up my bike, cycle for a mile and three quarters, have a puncture and then have to walk for a quarter of a mile. How far have I gone altogether?

3 Write a fraction that is bigger than $\frac{2}{3}$ and smaller than $\frac{5}{7}$

4 I have four and a half kilograms of potatoes and I use seven-eighths of a kilogram in a stew. How many kilograms do I have left?

5 I have a jar with $1\frac{3}{8}$ kg of jam in it and another jar with $1\frac{2}{5}$ kg of jam in it. How much jam do I have in the two jars altogether?

6 A snail is crawling up a window. First he crawls up $5\frac{1}{10}$ centimetres and then he slips down $1\frac{2}{3}$ centimetres. What is the total distance that he travelled before he reached his final position?

7 In a race, athletes have to run $4\frac{1}{4}$ miles, cycle $5\frac{3}{5}$ miles and swim $\frac{2}{3}$ miles. What is the total length of the race?

8 Last year I was four and a half feet tall. I have grown two-thirds of a foot in a year. How tall am I now?

9 We are on a $10\frac{1}{2}$ mile walk. We walked $4\frac{2}{3}$ of a mile before lunch and then walked $3\frac{2}{5}$ miles after lunch. How far have we walked in total and how much further do we have to go?

10 I have two boxes, one is $5\frac{1}{2}$ cm tall and the other is $10\frac{2}{5}$ cm tall.

(a) If I put them on top of each other what is their total height?

(b) I put them on a top shelf that is $17\frac{1}{4}$ cm below the ceiling. How big is the gap above them?

Exercise 7.8: Summary exercise

1 Copy these equations and fill in the missing numbers to make equivalent fractions.

(a) $\dfrac{2}{3}=\dfrac{\square}{36}$

(c) $\dfrac{3}{4}=\dfrac{9}{\square}$

(b) $\dfrac{5}{\square}=\dfrac{20}{80}$

(d) $\dfrac{\square}{8}=\dfrac{15}{24}$

2 Copy each pair of fractions and write < or > between them.

(a) $\dfrac{2}{5} \square \dfrac{4}{15}$

(b) $\dfrac{7}{4} \square 1\frac{5}{6}$

(c) $\dfrac{23}{10} \square 2\frac{4}{15}$

3 Write the fractions in each set in order, smallest first.

(a) $\dfrac{3}{5}, \dfrac{7}{15}, \dfrac{2}{3}$

(b) $\dfrac{3}{4}, \dfrac{5}{6}, \dfrac{7}{9}$

(c) $2\frac{6}{7}, \dfrac{17}{6}, 2\frac{2}{3}$

4 Add or subtract these numbers, giving your answers as fractions in their lowest terms. If the answer is an improper fraction, write it as a mixed number.

(a) $\dfrac{3}{5}+\dfrac{4}{15}$

(d) $3\frac{5}{6}+2\frac{3}{8}$

(b) $\dfrac{3}{4}-\dfrac{5}{12}$

(e) $3\frac{7}{15}-\dfrac{4}{5}$

(c) $2\frac{5}{6}+1\frac{5}{18}$

(f) $4\frac{4}{15}-1\frac{7}{12}$

5 Copy each sequence and fill in the missing terms.

(a) $3\frac{5}{12}$, $3\frac{7}{12}$, $3\frac{3}{4}$, $3\frac{11}{12}$, ..., ...

(b) $4\frac{3}{16}$ ..., $4\frac{13}{16}$ ..., $5\frac{7}{16}$, $5\frac{3}{4}$

6 Write a fraction that is bigger than $\frac{7}{10}$ and smaller than $\frac{11}{15}$

7 There are 20 pupils in the class and 12 of them are girls. What fraction of the class is this?

8 We have to drive 144 miles to our summer holiday cottage. By lunch time we had covered 108 miles. What fraction of the total distance have we left to travel?

9 It is three and three-quarter miles from Puddleton to Mudford. I have walked for one and four-fifths miles. How much further do I have to go?

10 We need ten and a half kilograms of sand. We have one bag with $4\frac{1}{4}$ kg in it and another with $3\frac{5}{6}$ kg.

(a) How much sand do we have?

(b) How much more do we need?

8 Ratio and proportion

⇨ Ratio in patterns

When you worked on number investigations, you considered patterns that grew in steps.

Look at this pattern.

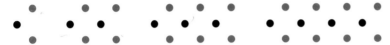

Pattern 1 has 1 black dot and 2 red dots.

Pattern 2 has 2 black dots and 4 red dots.

Pattern 3 has 3 black dots and 6 red dots.

Pattern 4 has 4 black dots and 8 red dots.

Altogether, there are always 2 red dots for every black dot.

Because there is always one black dot for every two red dots in this pattern, you can say that the **ratio** of black dots to red dots is 1 to 2

You write this as:

black : red = 1 : 2

Note that the **order** of the numbers is very important.

You could also write:

red : black = 2 : 1

Look at this pattern, drawn in an exercise book.

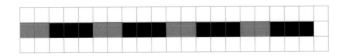

In this pattern there are two red squares for every three black squares.

red : black = 2 : 3

Like fractions, ratios must be written in their lowest terms.

In the pattern there are 8 red squares and 12 black squares in total.

So red : black = 8 : 12

Both 8 and 12 can be divided by 4 and therefore the correct ratio is:

red : black = 2 : 3

Exercise 8.1

1 Write each ratio in its lowest terms.

(a) 5 : 10   (b) 3 : 9   (c) 2 : 8   (d) 12 : 18

Copy each of these patterns into your exercise book. Write down the ratio of **red** squares to **black** squares for each one.

2

3

4

5

Copy each of these patterns into your exercise book. Write down the ratio of **black** squares to **red** squares for each one.

6

7

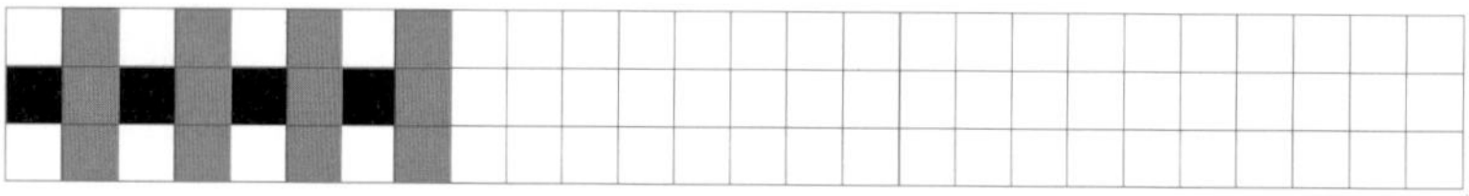

8

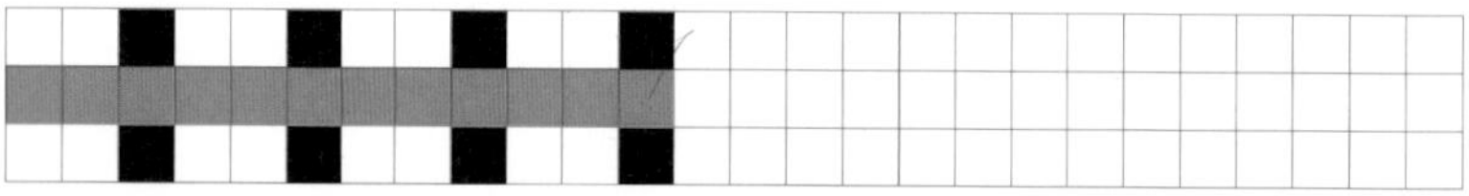

9

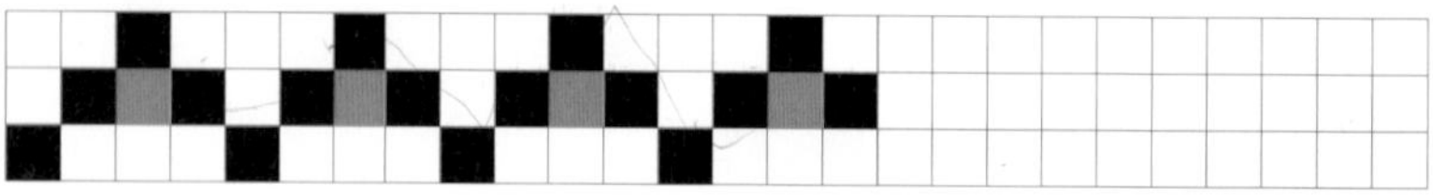

10 Draw some patterns of your own in a strip of squares. Your strip may be a maximum of three squares high.

(a) Draw a pattern with a ratio of black squares to red squares of 2 : 3

(b) Draw a pattern with a ratio of black squares to red squares of 5 : 1

(c) Draw two patterns with the ratio of black squares to red squares 4 : 1

⇨ Finding the ratio

You use ratio in all sorts of situations, for example, when you have a recipe and must keep the ingredients in the same proportions.

Example:

For a recipe for fruit salad you need 2 bananas for every 16 grapes.

What is the ratio of bananas to grapes?

Ratio of bananas to grapes = 2 : 16

$$= 1 : 8$$

Exercise 8.2

Remember to write each ratio in its simplest form by dividing by a highest common factor, just as you do when simplifying fractions.

1 In a bunch of flowers there are 4 roses and 12 carnations. What is the ratio of roses to carnations?

2 In a block of flats there are 16 doors and 48 windows. What is the ratio of doors to windows?

3 On a farm there are 16 cows and 24 chickens. What is the ratio of cows to chickens?

4 In a school there are 120 girls and 180 boys. What is the ratio of girls to boys?

5 In the junior dining room there are 12 tables and 24 benches. What is the ratio of tables to benches?

6 In the senior dining room there are 9 tables and 108 chairs. What is the ratio of tables to chairs?

7 In a box of chocolates there are 6 white chocolates and 15 dark chocolates. What is the ratio of white chocolates to dark chocolates?

8 In one week we study mathematics for 240 minutes and English for 300 minutes. What is the ratio of the time we study English to the time we study mathematics?

9 In the science laboratory there are 12 test-tube racks and 72 test tubes. What is the ratio of test-tube racks to test tubes?

10 We are making table decorations. In total we have 14 red balls and 21 white balls.

(a) What is the ratio of white balls to red balls?

(b) What is the ratio of white balls to the total number of balls?

11 The school is decorated with bunches of balloons. In total there are 240 blue balloons and 180 yellow balloons.

(a) What is the ratio of blue balloons to yellow balloons?

(b) If we have the maximum number of bunches, all with the same ratio of yellow to blue, how many balloons are there in a bunch?

⇨ Working with ratio

Consider this problem. You are making Christmas decorations with red and white balls. You have 14 red balls and 21 white balls. If every decoration has the same number of red and white balls how many decorations can you make?

In question 10 you had 14 red balls and 21 white balls.

The ratio of red to white = 14 : 21

$$= 2 : 3$$

We simplified the ratio by dividing both parts by 7

Therefore there can be 7 groups of red and white balls in total.

Consider 108 black tiles and 144 white tiles.

The ratio of black to white = 108 : 144

$$= 3 : 4$$

There must be 12 black tiles in each pattern.

The ratio of black to white = 3 : 4

$$= 12 : \boxed{}$$

You need to multiply 3 by 4 to get 12 and so you need to multiply 4 by 4 as well, to get the number in the box, which is 16

Write it like this.

The ratio of black to white = 3 : 4

$$= 12 : \boxed{} \qquad 12 \div 3 = 4$$
$$= 12 : 16 \qquad\quad 4 \times 4 = 16$$

Exercise 8.3

1 For each bunch of flowers a florist uses 4 roses and 12 carnations.

   (a) What is the ratio of roses to carnations?

   (b) If she uses 48 roses, how many carnations does she need?

   (c) How many bunches does she make?

2 For a school trip the ratio of teachers to pupils has to be 1 : 8

   (a) If they take 56 pupils, how many teachers must go as well?

   (b) If 4 teachers go on a trip, what is the maximum number of pupils that could go with them?

3 In science, Brian mixes 12 mg of chemical A with 18 mg of chemical B.

   (a) What is the ratio of chemical A to chemical B?

   (b) If he has 144 mg in total of chemical A, how many mixtures can he make?

   (c) How much chemical B will he need in total?

4 In a box of chocolates, there are 8 white chocolates and 20 dark chocolates.

(a) What is the ratio of white chocolates to dark chocolates?

(b) If the factory has made 240 white chocolates, how many boxes can they fill?

(c) How many dark chocolates will they need?

5 A balloon company delivers a total of 240 blue and red balloons.

(a) If 108 are blue, what is the ratio of blue balloons to red balloons?

(b) If I have to make identical bunches, how many balloons will there be in each bunch?

(c) How many bunches can I make?

6 24 pupils can sit comfortably on 6 benches.

(a) What is the ratio of pupils to benches?

(b) If the school has 32 benches, how many pupils can comfortably sit on them?

(c) If the school needs to seat 100 pupils on benches, how many benches will they need?

⇨ Using ratio

In the last two questions you had to answer questions after you had found the ratio. To do this you have to consider the number of parts in the total mix.

Look at this in more detail.

1 First find the total number of parts.

2 Work out the amount in one part.

3 Work out the amounts in all the parts.

4 Answer the question.

It is good practice to work out the amounts in all the parts and then to answer the question, as this means that you check that all your parts add up to the whole, and that you double check what the question is asking.

Example:

Lime-juice cordial is mixed with water in the ratio 1 : 3

How much cordial will I need to make 1 litre of lime-juice drink?

The ratio of lime-juice cordial to water to the total in the drink

Ratio   cordial : water : total = 1 : 3 : 4

$$= \boxed{} : \boxed{} : 1000 \qquad \text{1 litre} = 1000 \text{ ml}$$

$$= 250 : 750 : 1000 \qquad 1000 \div 4 = 250 \text{ and}$$
$$250 \times 3 = 750$$

I will need 250 ml of cordial.

Exercise 8.4

1 In a bunch of flowers the ratio of pink flowers to white flowers is 5 : 2

There are 21 flowers in the bunch.

(a) How many pink flowers are there in the bunch?

(b) How many white flowers are there in the bunch?

2 In a box of chocolates the ratio of white chocolates to dark chocolates is 2 : 3

There are 20 chocolates in a box.

(a) How many white chocolates are there in a box?

(b) How many dark chocolates are there in a box?

3 We are making table decorations from silver and gold balls. The ratio of silver balls to gold balls is 3 : 2

(a) In total, we have 600 balls. How many silver balls and how many gold balls do we have?

(b) One decoration has 4 gold balls. How many silver balls will it have?

(c) How many decorations do we make, in total?

4 On a school trip the ratio of teachers to pupils is 1 : 7
  The bus has 56 seats.

  (a) How many pupils can go on the bus?

  (b) How many teachers must go on the bus?

  (c) If 63 pupils want to go on a trip a bigger bus will need to
      be booked. How many seats will the bigger bus need?

5 Sand and cement are mixed in the ratio 4 : 1

  (a) If a builder has 12 kg of sand, how much cement will he
      need?

  (b) How much dry mixture will he have in total?

  (c) He needs to add water to the dry mix. The ratio of dry mix
      to water is 5 : 1. How much water will he need?

6 Fertiliser is diluted with water in the ratio 2 : 7

  (a) If I need 4.5 litres in total, how much water and how much
      fertiliser will I use?

  (b) A bottle of fertiliser concentrate contains 350 ml. What
      total volume of mixture can I make up?

7 To make an orange drink you add orange squash concentrate
  to water in the ratio 1 : 4

  (a) How much water do I need, if one bottle of concentrate
      contains 500 ml?

  (b) I want to make 5 litres of orange squash. How much
      concentrate will I need?

8 In a pattern, the ratio of black tiles to white tiles is 3 : 5

  (a) In total, we have 400 tiles. How many black tiles and how
      many white tiles do we have?

  (b) If we need 25 white tiles for one pattern, how many
      patterns can we make?

9 I have to draw rectangles with a width to length ratio
  of 2 : 3 and then compare their perimeters.

  (a) Write down the ratio   width : length : perimeter

(b) If the width of one rectangle is 2.5 cm, what is the length?

(c) If the length of one rectangle is 6 cm, what is the width?

(d) Compare the perimeters of the two rectangles. What do you notice?

10 I have to draw rectangles with a width to length ratio of 1 : 4 and then compare their areas.

(a) Write down the ratio   width : length.

(b) If the width of one rectangle is 1.5 cm, what is the length?

(c) If the length of one rectangle is 8 cm, what is the width?

(d) Compare the areas of the two rectangles. What do you notice?

⇨ Proportion

Sometimes the simplest way to solve problems is by comparing the quantities.

If a car is travelling at 30 miles per hour then:

- in 1 hour the car will travel 30 miles

- in 3 hours the car will travel 90 miles.

You could work this out simply by multiplying both the number of hours and the distance by three.

This method is useful in many situations, particularly when you have a recipe and must keep ingredients in the same proportion.

Examples:

This is the recipe for 8 rock cakes.

> 250 g self-raising flour
>
> 125 g sugar
>
> 125 g butter
>
> 1 egg

(i) I want to make 40 rock cakes. How much flour will I need?

For 8 rock cakes I need 250 g flour.

For 1 rock cake I need 250 ÷ 8 g flour.

For 40 rock cakes I need 40 × 250 ÷ 8 g flour.

$$= 250 × 5 \qquad (40 ÷ 8 = 5)$$

$$= 1250 \text{ g flour}$$

(ii) I have 4 eggs. How many rock cakes can I make?

With 1 egg I can make 8 rock cakes.

With 4 eggs I can make 8 × 4 rock cakes.

$$= 32$$

I can make 32 rock cakes.

Exercise 8.5

1 These are the ingredients in a recipe to make 10 scones.

> 240 g self-raising flour
>
> 60 g butter
>
> 30 g caster sugar
>
> 1 free-range egg, beaten

(a) I want to make 30 scones. How much sugar will I need?

(b) I have 4 eggs. How many scones can I make?

(c) I only have 48 g of butter. How many scones can I make?

2 These are the ingredients in a recipe for pasta sauce that will serve 4 people.

> 1 kg chopped tomatoes
>
> 350 g beef mince
>
> 2 onions
>
> 60 g tomato concentrate

(a) I have to feed 6 people. How much chopped tomato will I need?

(b) I have 6 onions. How many people can I serve?

(c) I want to serve 10 people. How much beef mince will I need?

3 These are the ingredients in a recipe for 12 chocolate crispies.

> 60 g unsalted butter
>
> 3 tablespoons golden syrup
>
> 1 × 100 g bar milk or dark chocolate
>
> 90 g crispie cereal

(a) We are going to make 240 chocolate crispies to sell at the school fete. Write out the ingredients that we need to buy.

(b) I practise at home but only have 40 g of butter. How many chocolate crispies can I make?

(c) If I make chocolate crispies with 150 g of chocolate, how many tablespoons of golden syrup will I need?

4 The instructions on a fertiliser bottle say that 200 ml of concentrated fertiliser should be mixed with 10 litres of water.

(a) How much water will I need for 75 ml of concentrate?

(b) How much concentrate will I need to mix with 5 litres of water?

5 Modelling clay can be made by mixing 2 parts water with 3 parts cornflour.

(a) I have 600 mg of cornflour. How much water do I need? 400mg

(b) I want to make 1.5 kg of modelling clay. How much cornflour do I need? 400g

6 This is a recipe from an old book.

To make gunpowder you need – seventy-five parts saltpetre finely ground, fifteen parts charcoal and ten parts sulphur. All ingredients must be fine ground separately. This can be done by hand using a pestle and mortar. Never mix all three ingredients before grinding unless you want to turn your mortar into a canon and blow your hand off!

(a) If you wanted to make 1 kg of gunpowder how much would you need of:

(i) saltpetre   **(ii)** charcoal   **(iii)** sulphur?

(b) If you had 300 g of saltpetre, how much gunpowder could you make?

7 You can make glue by mixing 7 parts flour, 10 parts water, 3 parts sugar and a dash of vinegar.

(a) I want to make a kilogram of glue. How much will I need of:

(i) flour   **(ii)** water   **(iii)** sugar?

(b) I have 450 g of sugar. How much flour and water will I need?

(c) I have 210 g of flour. How much water and sugar will I need?

⇨ Equivalent measures

You can use the same method to compare equivalent measures, particularly when you are comparing imperial and metric quantities, or currencies.

Example:

A mass of 5 kg is equivalent to 11 lbs.

What mass, in kilograms, is equivalent to 44 lbs?

11 lbs is equivalent to 5 kg

1 lb is equivalent to $\frac{5}{11}$ kg

44 lbs is equivalent to $\frac{5}{11} \times 44$

$= 20$ kg

20 kg is equivalent to 44 lbs.

Exercise 8.6

1 A volume of 20 gallons is equivalent to 90 litres.

(a) How many gallons will there be in a 45-litre barrel? *10 gallons*

(b) How many litres are there in a 4-gallon bucket? *18 lts*

2 £10 is worth US$16

(a) How much is £45 worth, in US dollars? *$70*

(b) In America a new smart phone costs US$240 *£150*
What is this in English pounds?

$$16\overline{)2\overset{1}{4}\overset{5}{8}}$$

3 £20 is worth €30

(a) A coat costs €100 in France. What is that in English pounds?

(b) I have £25 to take on the ski trip. How many euros is that worth?

4 A distance of 5 miles is equivalent to 8 km.

(a) What distance, in kilometres, is equivalent to:

(i) 10 miles *16 Km*

(ii) 50 miles *80 Km*

(iii) 75 miles *120 Km*

(iv) 120 miles? *182*

(b) What distance, in miles, is equivalent to:

(i) 16 kilometres *10*

(ii) 48 kilometres *30*

(iii) 100 kilometres

(iv) 240 kilometres?

5 A mass of 10 kg is equivalent to 22 lbs (pounds weight).

(a) What weight, in pounds, is equivalent to:

(i) 5 kg *11 Lbs* (ii) 12 kg?

(b) What mass, in kilograms, is equivalent to:

(i) 2 lbs (ii) 10 lbs?

6 A length of 1 m is equivalent to 39 inches, or 3 feet and 3 inches.

   (a) What length, in feet and inches, is equivalent to:

      (i) 5 m   **(ii)** 10 m?

   (b) What length, in metres, is equivalent to 13 feet?

⇨ Ratio and scale

When you are planning where to put plants in a garden, or furniture in a room, it can be helpful to use a plan. When you look at plans and maps you may not realise that you are actually using ratio.

Plans and maps are drawn to **scale**, and the scale is a ratio. This may be expressed as 1 cm for every 10 m, 1 cm to 50 cm, or 1 : 100

You may also use scale drawings and scale models to help you think about larger places or objects.

When using scale, first make sure that you know which is the scale drawing and which is the original, then check the ratio.

Example:

Lewis has a model of a sports car made to a scale of 1 : 50

  (i) If the model is 3 cm tall, how high is the actual sports car?

  (ii) If the actual sports car is 3 m long, how long is the model?

  (i) Scale : real = 1 : 50   Write the scale.

             = 1 cm : 50 cm   Write the units.

             = 3 cm : ☐ cm   Write the known dimensions.

             = 3 cm : 150 cm   Scale up or down.

     The real car is 1.5 m high.   Write the answer with the correct units.

(ii) Scale : real = 1 : 50   Write the scale.

             = 1 cm : 50 cm   Write the units.

             = ☐ cm : 300 cm   Write the known dimensions.

             = 6 cm : 300 cm   Scale up or down.

   The model car is 6 cm long.   Write the answer with the correct units.

Exercise 8.7

1 A model train set is made to a scale of 1 : 100

  (a) What does 1 cm on the model train set represent?

  (b) If the model is 15 cm long, how long is the actual train?

  (c) If the actual train has a height of 5 m, what is the height of the model?

2 A model train set is made to a scale of 1 : 100
Thomas is making some buildings to add to the landscape.

  (a) He wants to make a station house that is a model of a real station house 12 m long, 6 m wide and 5 m tall. What are the dimensions of the model?

  (b) He adds a model of a red telephone box. If a real telephone box is 1 m by 1 m by 2.5 m, what are the dimensions of the model?

3 Peter has a remote-controlled model racing car made to a scale of 1 : 10

  (a) If the model is 43 cm long, how long is the actual racing car, in metres?

  (b) If the racing car is 2.3 m wide, how wide is the model, in centimetres?

4 I have a dolls' house that is a model of our house, on a scale of 1 : 10

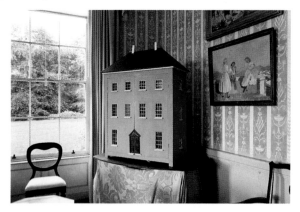

(a) What length does 1 cm on the dolls' house represent?

(b) If the model is 1 m high, what is the height of our house?

(c) If our house is 7 m wide, what is the width of the dolls' house?

(d) I am making furniture for my dolls' house. Work out the scaled dimensions for:

   (i) a bed 2 m by 60 cm by 50 cm

  (ii) a table 1.5 m by 90 cm by 90 cm

  (iii) a cupboard 1.8 m by 60 cm by 1 m

(e) I have made a model dog's kennel that is 6 cm by 8 cm by 9 cm.
How many model dog's kennel that is 6 cm by 8 cm by 9 cm.
What are the dimensions of the actual dog's kennel?

5 I have a model of the *Titanic* made to a scale of 1 : 1000

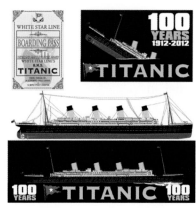

(a) What does 1 cm on the model represent?

(b) If the model is 27 cm long, how long was the *Titanic*?

(c) If the beam of the model is 2.8 cm, what was the beam of the *Titanic*?

(d) If the *Titanic* had a height of 32 m from the bottom of the keel to the height of the bridge, what is the height of the model?

(e) If the model has 4 funnels, how many funnels did the *Titanic* have?

6 I have a model Boeing 777 aeroplane made to a scale of 1 : 500

(a) What does 1 cm on the model represent?

(b) If the model is 12 cm long, how long is the actual aeroplane?

(c) If the aeroplane has a height of 18 m, what is the height of the model?

7 This is an engineer's drawing of a foot bridge drawn to a scale of 1 : 50

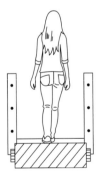

(a) What does 1 cm on the drawing represent?

(b) How wide is the bridge?

(c) How tall is the bridge?

(d) How tall is the pedestrian on the bridge?

(e) Measure your own height and width and draw a scale model of yourself that could be added to the drawing.

8 A room is in the shape of a rectangle 3 m by 4 m. Draw a plan of the room, to a scale of 1 : 20

   (a) Mark on your plan the position of a door, 60 cm wide, and a window 1 m wide.

   (b) Mark on your plan a bed measuring 2m by 60 cm and a table measuring 1 m by 50 cm.

Exercise 8.8: Summary exercise

This exercise will bring together what you have learnt.

1 Write each ratio in its lowest terms.

   (a) 2 : 6   **(b)** 10 : 25   **(c)** 16 : 6

2 Copy this pattern in your exercise book and write down the ratio of **red** squares to **black** squares.

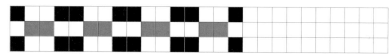

3 Copy this pattern in your exercise book and write down the ratio of **black** squares to **red** squares.

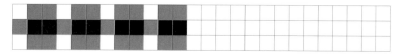

4 Draw a pattern in which the ratio of red squares to black squares is 2 : 3

5 In science, Louis mixes 15 mg of chemical X with 25 mg of chemical Y.

   (a) What is the ratio of chemical X to chemical Y?

   (b) If he has 90 mg, in total, of chemical X, how many mixtures can he make?

   (c) How much of chemical Y will he need, in total?

6 To make blackcurrant squash you mix concentrate to water in the ratio 1 : 3

 (a) How much water do I need, if one bottle of concentrate contains 500 ml?

 (b) I want to make 6 litres of blackcurrant squash. How much concentrate will I need?

7 These are the ingredients in a recipe that will make 20 biscuits.

> 250 g butter
>
> 150 g sugar
>
> 1 egg yolk
>
> 300 g flour

 (a) We are going to make 300 biscuits to sell at the school fete.

  (i) How much butter will we need to buy?

  (ii) How much sugar will we need to buy?

  (iii) How many eggs will we need to buy?

  (iv) How much flour will we need to buy?

 (b) I practise at home but only have 200 g of butter. How many biscuits can I make?

8 A volume of 12 fluid ounces is equivalent to 350 millilitres.

 (a) How many fluid ounces will there be in a half-litre bottle?

 (b) A recipe asks for 4 fluid ounces. What is this in millilitres?

9 A model train carriage is made to a scale of 1 : 50

 (a) If the model is 12 cm long, what is the length of the real carriage?

 (b) If the carriage is 4 m wide, what is the width of the model?

10 This map is drawn to a scale of 1 : 50 000

Each grid square represents 1 square kilometre

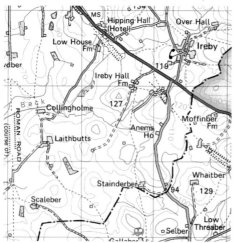

(a) What distance will 1 cm on the map represent?

(b) Measure the distance, in centimetres, from the telephone box in Ireby to the caravan site in the south.

(c) Calculate the true distance, in kilometres, from the telephone box in Ireby to the caravan site in the south.

(d) I am planning a walk from Collingholme to Ireby Hall Farm by road. What is the distance on the map in centimetres?

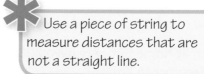

Use a piece of string to measure distances that are not a straight line.

(e) Calculate the true distance in kilometres from Collingholme to Ireby Hall Farm.

(f) We are staying at Moffinber Farm and want to bicycle for 2 km. How far is 2 km on the map?

(g) What is a suitable destination 2 km from Moffinber Farm by road?

9 Congruent and similar shapes

⇨ Congruent shapes

Two objects or shapes are **congruent** to each other if they are exactly the same shape and the same size.

Look at these five triangles.

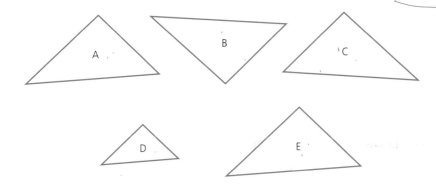

Triangles A and E are congruent because they are exactly the same shape and size.

Triangle D is not congruent to any other triangle because, although it is the same shape as all the others, it is smaller.

Are triangles B and C congruent to any of the other triangles?

If you trace over B and C with tracing paper, then turn your tracings over as well as around, you will see that they are both the same shape and the same size as triangles A and E so yes, they are congruent to triangles A and E.

Exercise 9.1

Use tracing paper to help you with this exercise.

1 Look at these triangles.

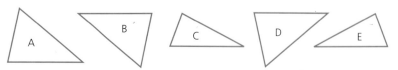

(a) Which triangles are congruent to triangle A? *B*

(b) Which triangles are congruent to triangle C? *E*

2 Look at these triangles.

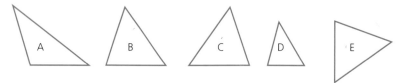

(a) Which, if any, are congruent to triangle A? *None*

(b) Which other triangles are congruent to each other?

3 Which of these quadrilaterals are congruent to each other?

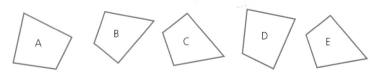

4 Which of these quadrilaterals are congruent to each other?

5 Which of these shapes are congruent to each other?

Similar shapes

Look at these two triangles.

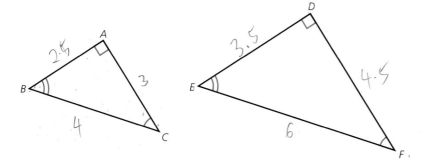

Their corresponding angles are the same size but the lengths of their sides are different.

If two objects or shapes are exactly the same shape but **not** the same size, they are **similar**.

Because they are similar, their sides are in the same **ratio**. This means that:

$AB : DE = AC : DF$

Also, $AB : DE = BC : EF$ and $AC : DF = BC : EF$

If you measure the sides of the triangles above, you will find that the ratio is 1 : 1.5 or 2 : 3

Ratios should be expressed in **integers**, rather than fractions or decimals.

Triangle *DEF* is bigger than triangle *ABC* by a **scale factor** of 1.5

Example:

In triangle ABC, $AB = 3$ cm, $AC = 4$ cm and $BC = 5$ cm. Triangle DEF is similar to triangle ABC and $DE = 4.5$ cm. Find the lengths of EF and DF.

First sketch the triangles and label the lengths of the sides that you know.

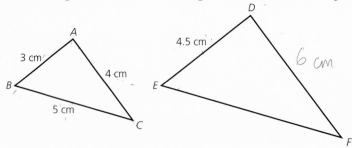

As you know that the triangles are similar, and therefore the sides are in the same ratio, you can write:

$AB : DE = 3 : 4.5$

$ = 6 : 9$

$ = 2 : 3$ Scale factor

$AC : DF = 2 : 3$ and $BC : EF = 2 : 3$ ×2.5 = 7.

$ = 4 : 6$ $= 5 : 7.5$

So $DF = 6$ cm and $EF = 7.5$ cm

29/8/2020

Exercise 9.2

1 These triangles are similar. Measure the lengths of their sides. Write down the ratio of their sides and the scale factor.

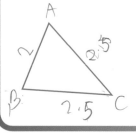

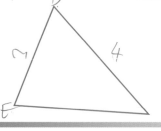

2 : 3
2 : 6
1 : 3

$AB : DE = 2 : 3$ What is scale factor?

$AC : DF = 2.5 : 3.5 \, 4$ $1 : 1.5$

$BC : EF = 2.5 : 4$

2 Which of these triangles are similar?

(a) Measure all the angles to find out.

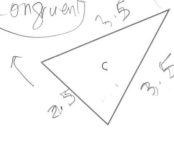
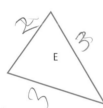
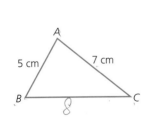

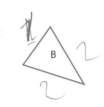
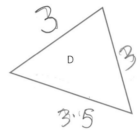

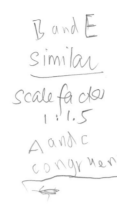

B and E
similar

scale factor
1 : 1.5

A and C
congruent

Congruent

Eg/
3 : 15
SF = $\frac{15}{3}$ = 1 : 5

Eg 4 : 24
$\frac{24}{4}$ = 6 = 1 : 6

(b) Now measure the lengths of the sides of the triangles that are similar.

Show that the lengths of the corresponding sides are in the same ratio.

3 Triangles *ABC* and *DEF* are similar.

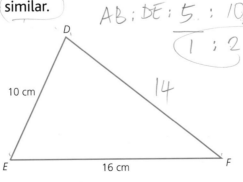

AB : DE : 5 : 10
1 : 2   1 : 2

$\frac{5}{10}$ = 2.

10 cm   14   16 cm

(a) What is the ratio of *AB* to *DE*? 1 : 2

(b) What is the length of: (i) *BC*   (ii) *DF*?
                              8 cm   14

4 Triangles *PQR* and *XYZ* are similar.

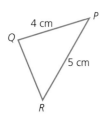 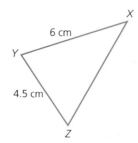

(a) What is the ratio of *PQ* to *XY*?

(b) What is the length of: (i) *XZ*   (ii) *QR*?

5 Triangles *LMN* and *RST* are similar.

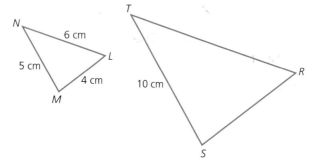

(a) What is the ratio *ST* : *MN*?

(b) What is the length of: (i) *RS*   (ii) *RT*?

6 These two quadrilaterals are similar.

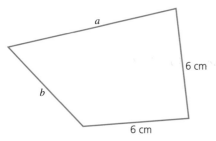 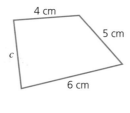

(a) What is the scale factor?

(b) Find the lengths of the sides labelled *a*, *b* and *c*.

7 Triangles *PQR* and *XYZ* are similar and
$PQ : XY = QR : YZ = PR : XZ$

$PQ = 5$ cm, $PR = 8$ cm, $XY = 7.5$ cm and $YZ = 9$ cm.

(a) Sketch the two triangles, marking the lengths of all the sides that you know.

(b) Calculate the ratio $PQ : XY$

(c) Calculate the lengths of the sides QR and XZ

(d) What are the ratios of the sides $PQ : QR : PR$ and $XY : YZ : XZ$?

8 Triangles *ABC* and *DEF* are similar.

$AB = 4.8$ cm, $BC = 3.6$ cm, $EF = 4.8$ cm and $DF = 8$ cm.

(a) Write down the ratios of the sides of the two triangles. Use the sequence of letters in the statement to guide you.

(b) Sketch the two triangles, marking all the sides that you know.

(c) Calculate the scale factor.

(d) Calculate the lengths of the sides AC and DE.

(e) What are the ratios of the sides $AB : BC : AC$ and $DE : EF : DF$?

9 In the rectangle *ABCD*, $AB = 5$ cm and $BC = 8$ cm. Rectangle *PQRS* is similar to rectangle *ABCD*, with $PQ = 10$ cm.

(a) What is the length of QR?

(b) What is the ratio of the area of *ABCD* to the area of *PQRS*?

10 Decimals

Whenever you use money, you are using decimals. The decimal point separates the pennies from the pounds.

Twelve pounds   £12

Twelve pence   either 12p or £0.12

⇨ Calculating with money

It is really useful to be able to make calculations about money in your head, without needing pencil and paper or a calculator. When you are shopping, you need to add the prices of what you buy, so you can check that you have enough money to pay for everything. You need to be able to work out your change, too.

If you are adding and subtracting, sometimes it is simpler just to think in pence.

> **Examples:**
>
> (i) Work out £1.25 + £3.42
>
> £1.25 + £3.42 = 125p + 342p
>
> = 467p
>
> = £4.67
>
> (ii) Calculate £5 − £2.65
>
> £5 − £2.65 = 500p − 265p
>
> = 235p
>
> = £2.35

When you are multiplying, remember to think carefully about the decimal point.

* Make sure you have the correct number of digits after the decimal point.

Example:

Work out £1.12 × 5

$$£1.12 × 5 = £1 × 5 + 12p × 5$$
$$= £5 + 60p$$
$$= £5.60$$

When you are dividing, sometimes you need to extend the numbers past the decimal point. Again, it can help to think in pennies.

Example:

Work out £4 ÷ 5

$$£4 ÷ 5 = 400p ÷ 5$$
$$= 80p$$
$$= £0.80$$

Exercise 10.1

Write down the answers.

1 £4 + £3.45

2 £5 − £3.12

3 £4.20 × 5

4 £10 − £6.35

5 £1.25 × 3

6 £10 ÷ 4

7 £8.50 − £7.90

Copy and complete these calculations.

8 £1.20 × ... = £6

9 £20 − ... = £10.01

10 48p × ... = £2.40

11 (£1.45 × 4) + ... = £10

12 ... ÷ 8 = £1.25

For these questions, first write down the calculation you need to do and then work out the answer.

13 What is the cost of 5 kilograms of potatoes if they cost £1.60 per kilogram?

14 My uncle gave us £10 to be shared equally among four of us. How much did we each get?

15 I bought four pens costing £1.45 each. How much change did I get from £10?

16 I have been saving £1.25 a week for 9 weeks. This week I spent £10.99 on a bunch of flowers for my mother's birthday. How much of my savings do I have left?

17 A shopkeeper is selling eggs at £2.25 for half a dozen. Is that better value than the farmer who is selling two dozen eggs for £8.40?

18 Copy and complete this shopping list.

2 kg of apples at £0.95 a kg   £.......

... pineapples at £1.60 each   £3.20

500g of grapes at £... a kg   £2.35
                                        _____
                        Total: _____

19 Copy and complete this shopping list.

... cauliflowers at 1.20 each   £4.80

3 kg of carrots at 95p a kg   £.......

5 kg potatoes at £... a kg   £4.15
                                        _____
                        Total: _____

20 Copy and complete this shopping list.

2 packets of biscuits at £1.35 each   £.......

... fairy cakes at 45p each   £3.60

5 pizzas at £... each   £16.75
                                        _____
                        Total: _____

⇨ Tenths, hundredths and thousandths

Because there are 100p in £1, one penny is one hundredth of a pound.

When you are working with decimals – and not with money – you may need more places of decimals. Extend the number columns to the right.

Hundreds	Tens	Units	•	tenths	hundredths	thousandths
100	10	1	•	$\frac{1}{10}$	$\frac{1}{100}$	$\frac{1}{1000}$

The place value in each column is 10 times smaller than that in the column to the left.

$$100 \div 10 = 10 \qquad 10 \div 10 = 1 \qquad 1 \div 10 = \frac{1}{10} \qquad \frac{1}{10} \div 10 = \frac{1}{100} \qquad \frac{1}{100} \div 10 = \frac{1}{1000}$$

Note that the decimal point sits between the whole numbers and the fractions. It does not have a column of its own.

The position of the digit gives its value.

Examples:

Write down the real value of the underlined digit in each number.

(i)

T	U	•	t	h	th
2	3	•	0	<u>9</u>	5

9 is in the hundredths column, so it is $\frac{9}{100}$

(ii)

T	U	•	t	h	th
2	3	•	0	9	<u>5</u>

5 is in the thousandths column, so it is $\frac{5}{1000}$

(iii)

T	U	•	t	h	th
2	3	•	<u>0</u>	9	5

0 is in the tenths column, so it is $\frac{0}{10} = 0$

It has no value.

➯ Multiplying and dividing by 10, 100 and 1000

When you multiply a number by 10, 100 or 1000 the digits move left, into different columns. You write zeros in any whole-number column that no longer has a digit in it.

H	T	U ·	t	h	th
		8			
	8	0			
		7 ·	3	5	
7	3	5			
		0 ·	0	5	
	5	0			

8 × 10

= 80 — The digits move 1 place to the left.

7·35 × 100

= 735 — The digits move 2 places to the left.

0·05 × 1000

= 50 — The digits move 3 places to the left.

When you divide a number by 10, 100 or 1000 the digits move right, into different columns. You write zeros in any column that no longer has a digit in it.

Th	H	T	U ·	t	h	th
		1	9			
			1 ·	9		
			2 ·	7		
			0 ·	0	2	7
3	7	5	0			
			3 ·	7	5	

19 ÷ 10

= 1·9 — The digits move 1 place to the right.

2·7 ÷ 100

= 0·027 — The digits move 2 places to the right.

3750 ÷ 1000

= 3·75 — The digits move 3 places to the right.

Exercise 10.2

First write down the calculation you need to do and then work out the answer.

Write your answer in words.

Examples:

(i) Multiply four-hundredths by 10

(ii) Divide thirty by a thousand.

(i) $0.04 \times 10 = 0.4$

Four tenths

(ii) $30 \div 1000 = 0.03$

3 hundredths

1 Multiply three-tenths by a hundred.

2 Divide four by ten.

3 Multiply two-thousandths by ten.

4 Divide five by a thousand.

5 Multiply six-hundredths by a thousand.

6 Divide twenty by a thousand.

7 Multiply nine-hundredths by a hundred.

8 Divide two-tenths by a hundred.

9 Multiply four-thousandths by a thousand.

10 Divide six hundred by a thousand.

Write down the answers.

11 3.14 × 10

12 0.43 ÷ 10

13 1.35 × 1000

14 215 ÷ 100

15 19.28 × 100

16 0.12 × 10

17 275 ÷ 1000

18 6.3 × 100

19 4.5 ÷ 100

20 7.2 × 1000

21 2.08 × 10

22 4.05 ÷ 10

23 0.604 × 1000

24 705 ÷ 100

25 10.04 × 100

26 0.302 × 10

27 801 ÷ 1000

28 3.05 × 100

29 4050 ÷ 100

30 7.002 × 1000

⇨ Rounding decimals

Think about 5 thousandths. You should recognise that this is a very small number, compared to a whole.

Sometimes you do not need to include all the decimal places in an answer. You can round to a smaller number of places or to a whole number.

Consider the calculation 5 ÷ 7

	U ·	t	h	th	tth
	0 ·	7	1	4	2
7	5 · ^{5}0	^{1}0	^{3}0	^{2}0	

Step 1: 7 into 5 will not go, carry the 5

Step 2: 7 into 50 is 7 remainder 1, carry the 1

Step 3: 7 into 10 is 1 remainder 3, carry the 3

Step 4: 7 into 30 is 4 remainder 2, carry the 2

Step 5: 7 into 20 is 2 remainder 6, carry the 6

You can see this calculation could go on for a long time, but you can round decimal numbers, just as you rounded whole numbers in Chapter 2

If the value of a digit falls below the halfway mark, round down.

If the value of a digit falls at or above the halfway mark, round up.

Look at 0.7142 on a number line.

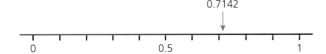

As 0.7142 is between 0 and 1 but more than 0.5, round up.

Then 0.7142 = 1 **to the nearest whole number**.

Now consider the first decimal place.

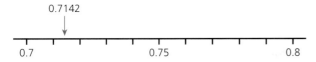

As 0.7142 is between 0.7 and 0.8 but less than 0.75, round down.

Then 0.7142 = 0.7 **to 1 decimal place** (1 d.p.).

Now consider the second decimal place.

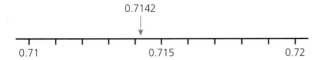

As 0.7142 is between 0.71 and 0.72 but less than 0.715, round down.

Then 0.7142 = 0.71 **to 2 decimal places** (2 d.p.).

Now consider the third decimal place.

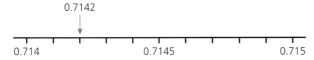

As 0.7142 is between 0.714 and 0.715 but less than 0.7145, round down.

Then 0.7142 = 0.714 **to 3 decimal places** (3 d.p.).

You can probably see a simple rule developing.

1 Count the number of decimal places that you need.

2 Look at the digit to the right.

3 If it is 5 or more, round your digit up.

4 If it is 4 or less, leave your digit as it is.

Examples:

(i)   Round 14.516 to:

  (a)   the nearest whole number

  (b)   1 decimal place

  (c)   2 decimal places.

  (a)   14.516 = 15 (to the n.w.n.)   As the digit in the first d.p. is 5 and is equal to 5

  (b)   14.516 = 14.5 (to 1 d.p.)   As the digit in the second d.p. is 1 and is less than 5

  (c)   14.516 = 4.52 (to 2 d.p.)   As the digit in the third d.p. is 6 and is more than 5

(ii)   Round 0.7098 to:

  (a)   1 decimal place

  (b)   2 decimal places

  (c)   3 decimal places.

  (a)   0.7098 = 0.7 (to 1 d.p.)   As the digit in the second d.p. is 0 and is less than 5

  (b)   0.7098 = 4.71 (to 2 d.p.)   As the digit in the third d.p. is 9 and is more than 5

  (c)   0.7098 = 4.710 (to 3 d.p.)   As the digit in the fourth d.p. is 8 and is more than 5 and therefore the 9 becomes 10

In this case, you need to write the 0 to show that you have rounded to 3 d.p.

Exercise 10.3

1 Round these to the nearest whole number.

   (a) 12.5   (c) 6.5

   (b) 1103.3   (d) 99.9

2 Round these to one decimal place.

   (a) 12.17   (c) 0.153

   (b) 2.345   (d) 13.67

3 Round these to two decimal places.

   (a) 0.275   (c) 16.145

   (b) 14.056   (d) 12.0083

4 Round these to three decimal places.

   (a) 0.0567   (c) 0.3002

   (b) 4.0505   (d) 19.9999

5 Round these to:

   (i)   one decimal place

   (ii) two decimal places

   (iii) three decimal places.

   (a) 3.7382   (b) 0.0438   (c) 11.0548   (d) 14.0993

6 What are the largest and smallest numbers, with two places of
   decimals, that give 27 when rounded to the nearest whole number?

7 What are the largest and smallest numbers, with two places of
   decimals, that could give 100 when rounded to the nearest whole
   number?

8 What are the largest and smallest numbers, with three places of
   decimals, that could give 4.6 when rounded to one decimal place?

9 What are the largest and smallest numbers, with four places of
   decimals, that could give 17.14 when rounded to two decimal
   places?

10 What are the largest and smallest numbers, with four places of
   decimals, that could give 0.400 when rounded to three decimal
   places?

⇨ Formal written calculations

When a calculation is too difficult for you to do mentally, you can write the numbers in a frame.

Addition and subtraction

When the numbers include decimals, it is important to write the decimal points in the correct place.

Examples:

(i) Add: $14.5 + 6007.8 + 0.412$

Th	H	T	U	t	h	th
		1	4	5		
6	0	0	7	8		
			0	4	1	2
6	0	2	2	7	1	2
		1	1			

Write each number so that the decimal points align with each other, like a neat row of buttons.

Add as normal.

(ii) Subtract: $12.3 - 5.715$

T	U	t	h	th
1	^{1}2	^{12}3	^{9}0	^{1}0
	5	7	1	5
	6	5	8	5

Write each number so that the decimal points align with each other, like a neat row of buttons.

Write zeros in the empty decimal columns.

Subtract as normal.

Exercise 10.4

Calculate the answers. Show all your working, including any carried numbers.

1 3.2 + 5.734

2 40.5 − 3.375

3 19 + 0.754 + 5.9

4 102 − 5.732

5 19.5 + 1004 + 0.595

6 300 − 10.745

7 17.15 + 6.456 + 105

8 909 − 10.375

9 19 + 9.9 + 199 + 0.199

10 34 − 3.107

Multiplication

Just as with addition and subtraction, the decimal points must align with each other.

Examples:

(i) Multiply: 5.137 × 6

Start with an estimate of 5 × 6

The answer is 30 and so you will need to have the tens column for your answer.

Multiply as normal.

Remember to write the decimal point in the answer row.

T	U	· t	h	th
	5	· 1	3	7
			×	6
3	0	· 8	2	2
3		2	4	

5.137 × 6 = 30.822

(ii) Multiply: 3.514 × 5

Estimate: 4 × 5 = 20

Multiply as normal.

Remember to write the decimal point in the answer row.

When you write your final answer out you do not need to write the last 0

T	U	· t	h	th
	3	· 5	1	4
			×	5
1	7	· 5	7	0
1	2		2	

3.514 × 5 = 17.57

Calculate the answers to these. Show all your working, including any carried numbers.

1 4.13×4

2 2.74×3

3 7.09×5

4 45.7×6

5 7.35×4

6 14.19×9

7 6.432×7

8 0.586×5

9 1.712×8

10 5.492×6

Multiplying by a two-digit number

When you need to multiply by a two-digit number that is a multiple of ten, remember to write the 0 in the answer line straight away.

Example:

Multiply: 3.28×40

Estimate: $3 \times 40 = 120$

H	Th	U	t	h
		3	2	8
		×	4	0
1	3	1	2	0
	1	3		

Add the zero here.

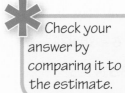

Check your answer by comparing it to the estimate.

$3.28 \times 40 = 131.2$

Exercise 10.6

1 3.14 × 20

2 5.42 × 30

3 1.94 × 50

4 3.26 × 40

5 4.32 × 60

6 24.72 × 80

7 35.72 × 70

8 5.427 × 90

9 8.342 × 60

10 19.54 × 50

When the two-digit number is not a multiple of ten, you can multiply in exactly the same way as you did in Chapter 3. You could multiply by factors or you could use long multiplication.

Examples:

(i) Multiply: 3.12 × 32

(ii) Multiply: 7.48 × 63

(i) 3.12 × 32

Estimate 3 × 30 = 90

32 = 4 × 8

So 3.12 × 32 = 3.12 × 4 × 8

= 12.48 × 8

= 99.84

3.12 × 32 = 99.84

H	T	U	·	t	h
	1	2	·	4	8
			×		8
	9	9	·	8	4
	1	3	6		

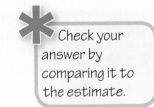

Check your answer by comparing it to the estimate.

(ii) 7.48 × 63

Estimate: 7 × 60 = 420

H	Th	U	·	t	h
		7	·	4	8
		×		6	3
	2	2₁	·	4₂	4
4	4₂	8₄	·	8	0
4	7	1	·	2	4
	1	1			

Set up the frame.

Write down what you are multiplying by in each row.

Multiply by 3, then by 60

× 3

The carried numbers from the × 3 are in this row.

× 60

The carried numbers from the × 60 are in this row.

The carried numbers from the addition are here.

7.48 × 63 = 471.24

Check your answer.

Exercise 10.7

1 5.12 × 24

2 3.42 × 18

3 4.73 × 32

4 1.05 × 54

5 3.63 × 45

6 35.72 × 84

7 42.17 × 27

8 0.924 × 49

9 45.65 × 72

10 40.59 × 56

11 6.14 × 29

12 2.45 × 37

13 3.53 × 19

14 6.37 × 53

15 2.73 × 41

16 4.645 × 74

17 12.09 × 89

18 0.736 × 59

19 36.82 × 47

20 60.12 × 73

Division

Just as before, keep the decimal points aligned with each other.

Examples:

(i) Divide: $9.24 \div 7$

	U	·	t	h
		1	· 3	2
7	9	· ^{2}2	^{1}4	

Step 1: $9 \div 7 = 1$, remainder 2. Write 1 in the answer line and carry the 2

Step 2: $22 \div 7 = 3$, remainder 1

Step 3: $14 \div 7 = 2$

$9.24 \div 7 = 1.32$

(ii) Divide: $2.716 \div 4$

	U	·	t	h	th
		0	· 6	7	9
4	2	· ^{2}7	^{3}1	^{3}6	

Step 1: $2 \div 4$ doesn't work, so look at $27 \div 4 = 6$ remainder 3. Write 6 in the answer line and carry 3

Step 2: $31 \div 4 = 7$, remainder 3. Write 7 in the answer line and carry 3

Step 3: $36 \div 4 = 9$. Write 9 in the answer line.

$2.716 \div 4 = 0.679$

You may have to extend the number with extra zeros and keep dividing.

Example:

Divide: $5.7 \div 8$

	U	·	t	h	th	tth
		0	· 7	1	2	5
8	5	· ^{5}7	^{1}0	^{2}0	^{4}0	

Step 1: $5 \div 8$ doesn't work, so look at $57 \div 8 = 7$ remainder 1. Write 7 in the answer line, add a 0 to 5.7 and carry 1

Step 2: $10 \div 8 = 1$, remainder 2. Write 1 in the answer line, add a 0 to 5.70 and carry 2

Step 3: $20 \div 8 = 2$ remainder 4. Write 2 in the answer line, add a 0 to 5.700 and carry 4

Step 4: $40 \div 8 = 5$. Write 5 in the answer line.

$5.7 \div 8 = 0.7125$

Calculate:

1 7.2 ÷ 2	**6** 8.68 ÷ 7
2 3.6 ÷ 3	**7** 10.8 ÷ 8
3 5.12 ÷ 4	**8** 6.93 ÷ 9
4 6.2 ÷ 5	**9** 31.2 ÷ 5
5 8.16 ÷ 6	**10** 4.29 ÷ 6
11 6.12 ÷ 9	**16** 3.48 ÷ 6
12 7.28 ÷ 7	**17** 8.33 ÷ 2
13 3.69 ÷ 6	**18** 1.27 ÷ 5
14 9.18 ÷ 4	**19** 6.42 ÷ 4
15 5.72 ÷ 8	**20** 1.96 ÷ 8

Remainders

These next questions will have remainders. Rather than continuing to add 0s, give your answers to 2 decimal places.

Example:

5.1 ÷ 7

	U	·	t	h	th	
	0	·	7	2	8	r4
7	5	·	⁵1	²0	⁶0	

Step 1: You cannot do 5 ÷ 7, so look at 51 ÷ 7 = 7, with 2 remainder. Write 7 in the answer line, add a 0 to 5.1 and carry 2

Step 2: 20 ÷ 7 = 2, remainder 6. Write 2 in the answer line, add a 0 to 5.10 and carry 6

Step 3: 60 ÷ 7 = 8 remainder 4. Write 8 in the answer line and remainder 4

5.1 ÷ 7 = 0.728...

Step 4: Write the answer followed by three dots to show that is not the complete answer.

= 0.73 (to 2 d.p.)

Then round to 2 d.p.

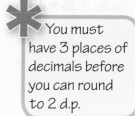

You must have 3 places of decimals before you can round to 2 d.p.

Exercise 10.9

1 $1.3 \div 8$

2 $5.62 \div 7$

3 $0.413 \div 6$

4 $2.06 \div 3$

5 $0.125 \div 9$

6 $1.03 \div 6$

7 $1.5 \div 7$

8 $0.3 \div 8$

9 $1.02 \div 9$

10 $5.8 \div 7$

Dividing by a two-digit number

As you saw in Chapter 3, there are different ways of dividing by two-digit numbers.

If the two-digit number is a multiple of 10, first divide by ten and then divide again.

Example:

Divide: $3.15 \div 90$

$$3.15 \div 90 = 3.15 \div 10 \div 9$$

$$= 0.315 \div 9$$

	U	.	t	h	th
		0	0	3	5
9		0	3	³1	⁴5

Step 1: $0 \div 9 = 0$, $3 \div 9 = 0$ r 3. Write the 0s, carry the 3

Step 2: $31 \div 9 = 3$ remainder 4

Step 3: $45 \div 9 = 5$

$$3.15 \div 90 = 0.035$$

For some two-digit numbers you can divide by factors.

Example:

Divide: $3.36 \div 48$

$$3.36 \div 48 = 3.36 \div 6 \div 8$$

$$= 0.56 \div 8$$

$$= 0.07$$

Finally, you can use long division.

Example:

$22.57 \div 37$ Estimate $20 \div 30 \approx \frac{2}{3} \approx 0.6$

		T	U	·	t	h
			0	·	6	1
3	7	2	2	·	5	7
		2	2	2		
					3	7
					3	7
					-	-

	3	7
	×	6
2	2	2
	4	

Divide.

Multiply.

Subtract and pull down.

Divide, multiply.

Subtract.

* Check your answer by comparing it to the estimate.

$22.57 \div 37 = 0.61$

Exercise 10.10

Complete these divisions. You may need to put in extra zeros.

1 $2.6 \div 20$

2 $6.4 \div 40$

3 $12.5 \div 50$

4 $9.6 \div 60$

5 $0.4 \div 80$

6 $0.15 \div 30$

7 $8.2 \div 20$

8 $12.2 \div 40$

9 $11.3 \div 50$

10 $14.7 \div 60$

Use factors for these divisions. You may need to put in extra zeros.

11 $1.44 \div 24$

12 $22.05 \div 35$

13 $19.84 \div 64$

14 $15.04 \div 32$

15 $8.37 \div 27$

16 $1.8 \div 36$

17 $2.1 \div 35$

18 $102.6 \div 54$

19 $20.25 \div 81$

20 $18.9 \div 42$

Use long division for these. Again, you may need to put in extra zeros.

21 $7.14 \div 17$ 26 $11.28 \div 47$

22 $9.28 \div 29$ 27 $19.84 \div 31$

23 $53.2 \div 38$ 28 $4.93 \div 34$

24 $49.3 \div 58$ 29 $18.02 \div 53$

25 $8.17 \div 43$ 30 $136.9 \div 37$

⇨ Problem solving

When solving problems involving decimals, write down the numbers carefully and make sure you show your calculations clearly. Use estimates to make sure the decimal point is in the correct place.

Many of the problems that you will have to solve involve distance, mass and volume.

Make sure you know the standard metric units.

Length

10 **millimetres** (mm) = 1 centimetre (cm)

100 **centimetres** (cm) = 1 metre (m)

1000 millimetres (mm) = 1 metre (m)

1000 **metres** (m) = 1 **kilometre** (km)

Mass (weight)

1000 **milligrams** (mg) = 1 gram (g)

1000 **grams** (g) = 1 kilogram (kg)

1000 **kilograms** (kg) = 1 **metric tonne** (t)

Capacity (volume)

There are two units of capacity.

1000 **millilitres** (ml) = 1 **litre** (l)

1 Three numbers add up to 1. If two of the numbers are 0.753 and 0.145, what is the third number?

2 Four numbers add up to 100. If three of the numbers are 24.45, 1.643 and 17.006, what is the fourth number?

3 I have 16 equal numbers that add up to 100
What are the numbers?

4 What is the number that is exactly halfway between 4.26 and 5.18?

5 I have to divide 3 kg of flour equally into 8 bowls. How many kilograms will there be in each bowl? What is this in grams?

6 I am laying a path of paving slabs. If one slab is 24.5 cm, how long a path will 24 slabs make? How far is this in metres?

7 At sports day, Jackie threw the javelin 55.4 metres. Johnnie beat him by 24 cm. What was Johnnie's throw, in metres?

8 I need 30 fencing panels to go round my garden. If one fence panel is 1.6 m long, what is the total length of my fence?

9 My little brother put a book that was 24 mm high on top of a box that was 34.6 cm high, on top of a chair that was 0.65 m high. What was the height of the top of the box?

10 A recipe asks for 2.4 kg of potatoes, 475 g of onions, 300 g of cream and 250 mg of paprika. What is the total mass, in grams?

11 I have a large container of lemonade and I pour it into 24 glasses so that each contains 0.18 of a litre. What is that in millilitres? How many litres were in my container?

12 I add 1.2 kg of flour to 0.35 kg of sugar and 0.25 kg of milk. I whisk it all up and pour it into 25 little pots. How many kg of the mix is in each pot? What is this in grams?

13 The teacher has a 55 kg bag of cement. She measures out 2.4 kg for each of the 19 pupils. How much cement is left in the bag?

14 I have run for 0.45 km, walked for 2.3 km and still have 425 m to go until I get home. What is the total distance, in kilometres?

Exercise 10.12: Summary exercise

1 Copy and complete these calculations.

   (a) £4.53 + ... = £8

   (b) £10 − ... = £3.42

   (c) £2.36 + 75p + ... = £5

   (d) ... − £5.63 = £4.19

2 Write down the calculation you are doing and then the answer. Write your answer in words.

   (a) Multiply three-tenths by a hundred.

   (b) Divide twenty by a thousand.

   (c) Multiply six-hundredths by a thousand.

3 Round these numbers to:

   (i) the nearest whole number

   (ii) one decimal place

   (iii) three decimal places.

   (a) 44.5132   (b) 1.4503   (c) 230.4572

4 Calculate the answers to these, showing all your working clearly.

   (a) $13.653 + 1.9 + 0.452$ (e) 3.65×20

   (b) $9.2 - 5.385$ (f) 4.18×24

   (c) 5.16×37 (g) $5.4 \div 72$

   (d) $7 \div 80$ (h) $68.15 \div 47$

5 Copy and complete this shopping list.

   6 eggs at £2.90 per dozen   £.......

   ... litres of milk at 80p each   £2.40

   5 lemons at ... each   £1.80
                                    _____
                          Total: _____

6 When I add three numbers and divide the result by three I get 1.4. If two of my numbers are 2.145 and 0.391, what is the third number?

7 I have to walk 2.4 km. I have now walked for 385 m, how many kilometres have I left to go? What is this in metres?

8 The school cook fills each of 80 glasses with 0.35 ml of pudding. How many litres of pudding did she start with?

9 In science we shared 4 kg of fertiliser among 25 of us. How many kilograms did we each get? What is this in grams?

10 It takes 80 full buckets of water to empty the school pond? If the pond contains 350 litres of water, what is the volume of 1 bucket?

 # Fractions, decimals and percentages

Look at this number line. It shows some fractions and decimals between 0 and 1. It also shows some percentages from 0% to 100%

You can see that $0.1 = \frac{1}{10} = 10\%$

$$0.5 = \frac{1}{2} = 50\%$$

$$0.75 = \frac{3}{4} = 75\%$$

Remember how you can convert between equivalent fractions, decimals and percentages.

⇨ Decimals to fractions

You know that the value of a digit is shown by the column in which it is written.

T	U	t	h	th
	1 •	2		
	1 •	0	3	
1	9 •	0	0	7
	0 •	0	0	9

1 and 2 tenths is written 1.2

1 and 3 hundredths is written 1.03

19 and 7 thousandths is written 19.007

9 thousandths is written 0.009

Therefore $1.2 = 1\frac{2}{10}$ Divide 2 and 10 by 2 to simplify the fraction.

$$= 1\frac{1}{5}$$

Similarly $1.03 = 1\frac{3}{100}$ $19.007 = 19\frac{7}{1000}$ $0.009 = \frac{9}{1000}$

✳ When the numerator and denominator have common factors, you should simplify the fraction to its lowest terms.

Exercise 11.1

Write the decimals as fractions in their lowest terms.

1 1.4   6 12.025

2 0.04   7 0.005

3 10.25   8 1.37

4 250.3   9 1.375

5 11.125   10 1.372

⇨ Fractions to decimals

You should know these fractions and their decimal equivalents.

$$\frac{1}{2} = 0.5 \quad \frac{1}{4} = 0.25 \quad \frac{3}{4} = 0.75$$

You can change a fraction to a decimal by:

● finding the equivalent fraction with a denominator that is 10, 100 or 1000

$$\frac{7}{20} = \frac{35}{100}$$
$$= 0.35$$

● dividing the numerator (top number) by the denominator (bottom number).

$$\frac{3}{8} = 3 \div 8$$
$$= 0.375$$

	U	•	t	h	th
	0	•	3	7	5
8	3	•	³0	⁶0	⁴0

11 Fractions, decimals and percentages

Exercise 11.2

Write these fractions as decimals. Show your working.

1 $\frac{1}{8}$

2 $\frac{7}{8}$

3 $\frac{1}{5}$

4 $\frac{4}{5}$

5 $\frac{1}{20}$

6 $\frac{17}{20}$

7 $\frac{9}{40}$

8 $\frac{49}{50}$

9 $\frac{29}{50}$

10 $\frac{19}{40}$

⇨ Percentages to decimals and fractions

Percentage means 'out of a hundred'.

You can easily write any percentage as a fraction or a decimal.

A percentage that is over 100% is equivalent to a fraction or decimal that is greater than 1

$100\% = \frac{100}{100} = 1$

$$35\% = 0.35 = \frac{35}{100} = \frac{7}{20}$$

$$148\% = 1.48 = 1\frac{48}{100} = 1\frac{12}{25}$$

And in reverse:

$$0.3 = \frac{3}{10} = \frac{30}{100} = 30\%$$

$$1.56 = 1\frac{56}{100} = 156\%$$

And if it is a fraction:

$$\frac{3}{8} = 0.375 = 37.5\% \text{ or } = 37\frac{1}{2}\%$$

Note that the fractional part of the percentage can be written as a decimal fraction or as a normal fraction.

Exercise 11.3

Write these percentages as:

(a) decimals   **(b)** fractions in their lowest terms.

1 6%

2 13%

3 52%

4 65%

5 4%

6 172%

7 75%

8 15%

9 108%

10 54%

⇨ Decimals to percentages

A decimal is just a fraction out of 100, so to change a decimal to a percentage, just multiply by 100

Example:

Write 0.27 as a percentage.

$$0.27 = 0.27 \times 100\%$$

$$= 27\%$$

You must write the % sign in your answer to make the statement correct.

Exercise 11.4

Write these as percentages.

1 0.1

2 0.25

3 0.62

4 0.7

5 1.08

6 0.125

7 0.555

8 1.275

9 15.05

10 1.374

⇨ Fractions to percentages

A percentage is a fraction out of 100, so to change a fraction to a percentage, multiply by 100

Example:

Write $\frac{7}{20}$ as a percentage.

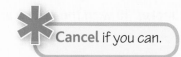
Cancel *if you can.*

$$\frac{7}{20} = \frac{7}{20} \times 100^5$$

$$= 35\%$$

Exercise 11.5

Write these as percentages.

1 $\frac{1}{2}$

2 $1\frac{1}{5}$

3 $\frac{1}{8}$

4 $\frac{1}{20}$

5 $\frac{1}{50}$

6 $1\frac{7}{10}$

7 $1\frac{9}{100}$

8 $1\frac{4}{25}$

9 $\frac{17}{40}$

10 $2\frac{1}{80}$

⇨ Recurring decimals

In Exercise 11.2, you found decimal equivalents of some fractions by dividing. All the fractions came out to two or three places of decimals.

Not all families of fractions turn out like this.

Example:

Write $\frac{1}{3}$ as a decimal.

	U	•	t	h	th	
	0	•	3	3	3	r1
3	1	•	¹0	¹0	¹0	

$$\frac{1}{3} = 1 \div 3$$

$$= 0.3333...$$

$$= 0.333 \text{ (to 3 d.p.)}$$

When you divide 10 by 3 the answer is 3 remainder 1

The remainder 1 then makes 10 with the next 0 and you get 3 remainder 1 again.

Note how the calculation is written. The fact that the answer continues to more decimal places is shown by the dots (...) after the last digit. Then the answer is rounded to 3 decimal places.

Another way of writing recurring decimals uses dots in a different way.

$0.3333... = 0.\dot{3}$

$0.191\ 919\ 19... = 0.\dot{1}\dot{9}$

$0.123\ 412\ 341\ 234 = 0.\dot{1}23\dot{4}$

The dots show the first and last digits of the repeating pattern.

$0.\dot{3}$ only has one dot, because it only has one recurring digit.

Example:

Write $\frac{4}{11}$ as a decimal.

$\frac{4}{11} = 4 \div 11$

$= 0.363\ 63...$

$= 0.364$ (to 3 d.p.)

$= 0.\dot{3}\dot{6}$

				0	•	3	6	3	6	r 4
1	1	4	•	⁴0	⁷0	⁴0	⁷0			

Exercise 11.6

Calculate the decimal equivalents for these families of fractions.

Remember to use the three dots to indicate that the answer continues to more decimal places, then round to 3 decimal places. In your final answer, use dots over the digits, to indicate the start and finish of the repeat.

1 (a) $\frac{1}{3}$ (b) $\frac{2}{3}$

2 (a) $\frac{1}{6}$ (b) $\frac{5}{6}$

Ask if you can use a calculator for this exercise, if you wish to.

3 (a) $\frac{1}{9}$ (b) $\frac{2}{9}$ (c) $\frac{3}{9}$ (d) $\frac{4}{9}$ (e) $\frac{5}{9}$

   (f) Predict the decimal equivalents to $\frac{6}{9}$, $\frac{7}{9}$ and $\frac{8}{9}$

4 Calculate the decimal values of the family of sevenths: $\frac{1}{7}, \frac{2}{7}, \ldots$

5 Calculate the decimal values of the family of elevenths: $\frac{1}{11}, \frac{2}{11}, \ldots$

6 Calculate the decimal values of the family of twelfths: $\frac{1}{12}, \frac{2}{12}, \ldots$

7 Calculate the first few decimal values of the family of 99ths: $\frac{1}{99}, \frac{2}{99}, \ldots$
Then suggest a general rule.

8 Investigate other fractions that you think may produce recurring decimals.

⇨ Writing recurring decimals as percentages

Consider a third as a decimal.

$$\frac{1}{3} = 1 \div 3$$

$$= 0.3333\ldots$$

Now multiply by 100 to change it to a percentage.

$$\frac{1}{3} = 33.3333\ldots \%$$

We know that $\frac{1}{3} = 0.3333\ldots$

And therefore

$$\frac{1}{3} = 33\frac{1}{3} \%$$

Example:

Write $\frac{4}{11}$ as a percentage.

$$\frac{4}{11} = 4 \div 11$$

$$= 0.363\ 63...$$

$$= 36.363\ 63...\%$$

$$= 36\frac{4}{11}\%$$

Exercise 11.7

Use your answers to Exercise 11.6 to write each of these fractions as a percentage with a fraction in the answer.

1 (a) $\frac{1}{3}$ (b) $\frac{2}{3}$

2 (a) $\frac{1}{6}$ (b) $\frac{5}{6}$

3 (a) $\frac{1}{9}$ (b) $\frac{2}{9}$ (c) $\frac{3}{9}$ (d) $\frac{4}{9}$ (e) $\frac{5}{9}$

(f) Predict the equivalent percentages for $\frac{6}{9}, \frac{7}{9}$ and $\frac{8}{9}$

4 Write the family of sevenths as percentages.

5 Write the family of elevenths as percentages.

6 Write the family of twelfths as percentages.

⇨ Ordering fractions, decimals and percentages

Sometimes you may have a mixture of percentages, fractions and decimals, and want to see which is the largest. The easiest way to do this is to write all the figures as **decimals**.

Example:

Write these in order, smallest first.

$\quad$ 33% $\quad \frac{3}{10}$ $\quad$ 0.35

$\quad$ 33% = 0.33 $\quad \frac{3}{10} = 0.3$ $\quad$ 0.35

Putting them in order:

$\quad \frac{3}{10}$, 33%, 0.35

Exercise 11.8

Write the numbers in each set in order, smallest first.

1 54% $\quad \frac{1}{2}$ $\quad$ 0.45

2 $\frac{1}{4}$ $\quad$ 26% $\quad$ 0.24

3 $\frac{3}{4}$ $\quad$ 72% $\quad$ 0.77

4 $\frac{1}{8}$ $\quad$ 0.13 $\quad$ 12%

5 22% $\quad \frac{1}{5}$ $\quad$ 0.19

6 175% $\quad 1\frac{4}{5}$ $\quad$ 1.82 $\quad 1\frac{7}{10}$

7 0.905 $\quad$ 89% $\quad \frac{9}{10}$ $\quad$ 0.92

8 $\frac{13}{20}$ $\quad$ 64% $\quad$ 0.66 $\quad \frac{3}{5}$

9 0.445 $\quad \frac{11}{25}$ $\quad$ 42% $\quad \frac{9}{20}$

10	$1\frac{5}{8}$	163%	1.62	$1\frac{3}{5}$	
11	77%	$\frac{15}{20}$	7.76	$\frac{7}{10}$	0.0777
12	1.303	131%	$1\frac{1}{3}$	1.03	$1\frac{3}{10}$
13	$\frac{8}{19}$	44%	0.444	$\frac{5}{11}$	45%
14	111%	$1\frac{1}{11}$	1.011	$1\frac{1}{9}$	1.01
15	$\frac{9}{11}$	81%	0.818	$\frac{8}{9}$	82%

Exercise 11.9: Summary exercise

1 Write these decimals as fractions.
   Remember to put them in their lowest terms.
   (a) 0.6   (b) 0.45   (c) 4.075   (d) 4.17

2 Write these fractions as decimals.
   (a) $\frac{9}{10}$ (b) $\frac{27}{100}$ (c) $\frac{13}{20}$ (d) $\frac{123}{500}$

3 Write these percentages as fractions in their lowest terms.
   (a) 40%   (b) 6%   (c) 82%   (d) 56%

4 Write these fractions as percentages.
   (a) $\frac{3}{4}$ (b) $\frac{4}{5}$ (c) $\frac{17}{25}$ (d) $\frac{1}{6}$

5 Write these percentages as decimals.
   (a) 3%   (b) 46%   (c) 19%   (d) $12\frac{1}{2}\%$

6 Write these decimals as percentages.
   (a) 0.6   (b) 0.03   (c) 1.952   (d) 2.875

7 Write the numbers in each set in order, smallest first.

(a) 0.65 $\frac{2}{3}$ 62%   0.606

(b) 0.207 $\frac{1}{4}$ 0.03 $27\frac{1}{2}\%$ $\frac{3}{11}$

8 Complete this table of equivalent fractions, decimals and percentages.

Fractions	Decimals	Percentages
	0.1	
$\frac{1}{5}$		
		30%
		40%
$\frac{1}{2}$		
	0.6	
		70%
$\frac{4}{5}$		
	0.9	
	0.25	
		75%
$\frac{1}{20}$		
	0.125	
$\frac{3}{8}$		
		$62\frac{1}{2}\%$
	0.875	
$\frac{1}{3}$		
	0.666 666...	

9 Make sure that you know the table above really well!

12 Finding fractions of an amount

In Chapter 7 you solved some problems by adding and subtracting fractions. Now think about this problem.

I have to divide a 10 metre length of rope into four equal pieces.

What is $\frac{1}{4}$ of 10 metres?

⇨ Finding a fraction of an amount: one part of a whole

To find one-quarter of 10 metres, you have to divide 10 metres by 4

$\frac{1}{4}$ of 10 metres = 10 ÷ 4

= 2.5 m

If you cannot do the division in your head you can write the calculation in a frame.

Examples:

(i) Find $\frac{1}{6}$ of £13.14

$\frac{1}{6}$ of £13.14 = £13.14 ÷ 6

= £2.19

			2 • 1	9	
	6	1	3 • ¹1	⁵4	

(ii) Find one-twelfth of 6 kg

$\frac{1}{12}$ of 6 kg = 6000 g ÷ 12

= 500 g

			5	0	0
1	2	6	0	0	0

✳ It can help to convert the quantity into smaller units.

Exercise 12.1

Calculate these amounts. If your answer is not exact, write it as a decimal.

1 $\frac{1}{4}$ of £10

2 $\frac{1}{2}$ of 5 m

3 $\frac{1}{3}$ of 6 km

4 $\frac{1}{5}$ of 400 g

5 $\frac{1}{10}$ of 3 kg

6 $\frac{1}{9}$ of £24.12

7 $\frac{1}{8}$ of 4 litres

8 $\frac{1}{6}$ of 3 m

9 $\frac{1}{5}$ of 2.4 kg

10 $\frac{1}{12}$ of 210 km

10 mm	= 1 cm
1000 mm	= 1 m
100 cm	= 1 m
1000 m	= 1 km
1000 mg	= 1 g
1000 g	= 1 kg
1000 ml	= 1 litre

11 It is 90 miles from London to Bath. Dad says we only have a quarter of the journey left to go. How many miles is that?

12 We each have $\frac{1}{8}$ of a 400 g bar of chocolate. How many grams is that?

13 A glass holds $\frac{1}{5}$ of a litre. How many millilitres is that?

14 A cook divides 2 kg of cake mix equally into 8 cake pans. How many grams are in each pan?

15 3 m of string is divided equally into twelfths. How long, in centimetres, is each piece?

16 I saved $\frac{1}{4}$ of my birthday money. If I was given £25, how much did I save?

17 2.4 litres is poured into 15 glasses. How many millilitres are there in each glass?

18 How many millimetres is $\frac{1}{5}$ of 4 cm?

19 A 9 cm line is divided into twelfths. How long, in millimetres, is each part?

20 For a science experiment we need some salt. A 2.4 kg bag is divided equally among 12 pupils. How many grams do we each get?

⇨ Finding a fraction of an amount: more than one part of a whole

If you are asked to find $\frac{3}{4}$ of an amount, first you need to find $\frac{1}{4}$ and then multiply the answer by 3 to find $\frac{3}{4}$

Examples:

(i) Find $\frac{2}{5}$ of 400 g

$\frac{1}{5}$ of 400 g = 400 g ÷ 5

= 80 g

$\frac{2}{5}$ of 400 g = 80 g × 2

= 160 g

(ii) Find $\frac{3}{4}$ of £1.40

$\frac{1}{4}$ of £1.40 = £1.40 ÷ 4

= £0.35

$\frac{3}{4}$ of £1.40 = £0.35 × 3

= £1.05

If you cannot do the division or multiplication in your head, you can write the calculation in a frame.

(iii) Find $\frac{5}{8}$ of 3 km

$\frac{1}{8}$ of 3 km = 3 km ÷ 8

= 0.375 km

$\frac{5}{8}$ of 3 km = 0.375 km × 5

= 1.875 km

			0 •	3	7	5	
		8	3 •	³0	⁶0	⁴0	

			0 •	3	7	5	
					×	5	
			1 •	8	7	5	
					3	2	

Exercise 12.2

Calculate each amount. If your answer is not exact, write it as a decimal.

1 $\frac{2}{3}$ of £12

2 $\frac{3}{5}$ of 20 kg

3 $\frac{3}{4}$ of 6 km

4 $\frac{3}{8}$ of 400 g

5 $\frac{7}{10}$ of 6 litres

6 $\frac{5}{6}$ of £21.42

7 $\frac{5}{8}$ of 2 km

8 $\frac{2}{5}$ of 3 litres

9 $\frac{5}{12}$ of 3.6 kg

10 $\frac{4}{9}$ of 2.7 km

11 $\frac{3}{4}$ of 3 kg

12 $\frac{9}{10}$ of £4.10

13 $\frac{3}{10}$ of 3 m

14 $\frac{5}{8}$ of 3 litres

15 $\frac{2}{3}$ of 4.5 kg

16 $\frac{2}{5}$ of £31

17 $\frac{3}{8}$ of 5 km

18 $\frac{5}{6}$ of 2.4 litres

19 $\frac{7}{9}$ of 2.7 g

20 $\frac{5}{12}$ of £1.80

21 In our class of 25 pupils, $\frac{3}{5}$ of are girls. How many girls are there?

22 How many grams are there in $\frac{2}{5}$ of 4 kg?

23 We are travelling from Birmingham to Bristol, a distance of 128 km. We stop for a cup of tea after travelling $\frac{5}{8}$ of the total distance. How far have we travelled by then?

24 I am going to spend $\frac{2}{5}$ of my total savings of £103.50 on some new clothes. How much do I spend?

25 A recipe needs $\frac{3}{8}$ of a litre of milk. How many millilitres is that?

26 I divide 800 g of flour so that I have $\frac{3}{5}$ and my sister has $\frac{2}{5}$ How much flour do we each get?

27 We have been given 2 kg of peppermints. I have $\frac{2}{5}$, I give $\frac{3}{8}$ to my sister and the rest to my brother. How many grams of peppermints do we each get?

⇨ Finding a fraction of an amount by cancelling

As the numbers get larger you will need to do more calculations.

It would be better if you could make the calculation simpler and then still do all the calculating mentally.

You know how to simplify fractions by dividing the numerator and the denominator by a common factor.

$$\frac{24}{100} = \frac{24 \div 4}{100 \div 4} = \frac{6}{25}$$

You can find a fraction of an amount in one calculation, if it is possible to divide by a common factor.

Example:

Find $\frac{5}{8}$ of 36

$\frac{5}{8}$ of $36 = \frac{5}{8} \times 36$ 4 is a common factor of 8 and 36

$\frac{5}{8}$ of $36 = \frac{5}{8} \times \overset{9}{\cancel{36}}$ Divide the top and bottom of the fraction by 4
$_{2}$

$ = \frac{45}{2}$ Multiply the top numbers.

$ = 22\frac{1}{2}$ Write the answer as a mixed number.

This process of reducing the size of the numbers in a calculation with fractions is called **cancelling**.

> If the amount has units you can leave the units out of the calculation but you must put them back in the final answer.

Example:

Find $\frac{4}{9}$ of 180 g.

$\frac{4}{9}$ of $180\text{ g} = \frac{4}{\cancel{9}} \times \overset{20}{\cancel{180}}$
$_{1}$

$ = 80\text{ g}$

Exercise 12.3

Calculate these amounts by cancelling. If your answer is not exact, write it as a fraction.

1 $\frac{2}{3}$ of 9

2 $\frac{3}{7}$ of 21

3 $\frac{3}{4}$ of 14

4 $\frac{3}{10}$ of 25

5 $\frac{5}{8}$ of 28

6 $\frac{5}{9}$ of 24

7 $\frac{2}{7}$ of 56

8 $\frac{7}{8}$ of 40

9 $\frac{3}{5}$ of 100

10 $\frac{3}{8}$ of 100

11 $\frac{5}{12}$ of 27

12 $\frac{4}{15}$ of 24

13 $\frac{7}{16}$ of 24

14 $\frac{11}{24}$ of 60

15 $\frac{7}{36}$ of 81

16 $\frac{13}{20}$ of 50

17 $\frac{24}{25}$ of 45

18 $\frac{7}{12}$ of 30

19 $\frac{9}{15}$ of 100

20 $\frac{15}{18}$ of 90

Sometimes you will need to change the units before you start the calculation.

Example:

Find $\frac{3}{5}$ of 6 kg

6 kg = 6000 g

$\frac{3}{5}$ of 6 kg = $\frac{3}{5} \times 6000$ g ^{1200}

= 3600 g or 3.6 kg

Exercise 12.4

Calculate these amounts by cancelling. If your answer is not exact, write it as a fraction.

1 $\frac{1}{6}$ of 3 kg

2 $\frac{7}{10}$ of 6 litres

3 $\frac{3}{5}$ of 2.5 km

4 $\frac{2}{3}$ of 600 g

5 $\frac{4}{7}$ of 1.4 kg

6 $\frac{2}{5}$ of £3

7 $\frac{3}{8}$ of 4 km

8 $\frac{4}{5}$ of 2 kg

9 $\frac{7}{9}$ of 270 g

10 $\frac{5}{12}$ of £3

11 $\frac{5}{8}$ of 6 kg

12 $\frac{7}{8}$ of 12 kg

13 $\frac{3}{10}$ of £15

14 $\frac{3}{5}$ of 2 litres

15 $\frac{8}{15}$ of £1.50

16 $\frac{5}{12}$ of £9

17 $\frac{7}{24}$ of 3 km

18 $\frac{3}{16}$ of £4

19 $\frac{3}{14}$ of 7 g

20 $\frac{7}{12}$ of 3 litres

⇨ Finding the original amount (1)

You can work out $\frac{1}{4}$ of 24 by dividing 24 by 4 and getting the answer 6

$$\frac{1}{4} \text{ of } 24 = 24 \div 4$$
$$= 6$$

What happens if the question is worded the other way?

If the 6 pupils who wear glasses are $\frac{1}{4}$ of the class, how many pupils are there in the class?

Write the calculation as the one above, but use n for the number of pupils.

$$\frac{1}{4} \text{ of } n = n \div 4$$
$$= 6$$

The important part is $n \div 4 = 6$

As multiplication is the opposite of division, you can solve the problem by multiplying.

$n = 6 \times 4 = 24$

Check your answer in the original question to make sure that you are correct.

Examples:

(i) If $\frac{1}{5}$ of an amount is 6, what is the total amount?

Let the total amount be n.

$n \div 5 = 6$

$n = 6 \times 5 = 30$

Check: $30 \div 5 = 6$

(ii) If $\frac{1}{3}$ of a length of rope is 15 m, how long is the whole rope?

Let the total amount be l.

$l \div 3 = 15$

$l = 15 \times 3 = 45$ m

Check: $45 \div 3 = 15$

Exercise 12.5

1 If $\frac{1}{4}$ of the total amount is 5, what is the total amount?

2 If $\frac{1}{5}$ of the whole is 4, what is the whole amount?

3 If $\frac{1}{3}$ of the total sum is 9, what is the total sum?

4 If $\frac{1}{7}$ of a number is 3, what is the number?

5 If $\frac{1}{6}$ of the total amount is 10, what is the total amount?

6 When a number is divided by 8 the answer is 4
What was the number?

7 When a number is divided by 12 the answer is 3
What was the number?

8 When a number is divided by 6 the answer is 12
What was the number?

9 When a number is divided by 5 the answer is 10
What was the number?

10 When a number is divided by 3 the answer is 9
What was the number?

11 We have travelled 105 km, and Mum says that we are half
way. What is the total length of our journey?

12 Our uncle shares a sum of money among four of us. If we each
receive £15, what was the total sum of money that my uncle
gave us?

13 Six cooks share a bag of flour equally. If each cook has 250 g
of flour, how much was in the whole bag?

14 If $\frac{1}{5}$ of a length of wood is 40 cm, how long is the whole
piece of wood?

15 One third of my class is going on the ski trip. If 9 pupils are
going skiing, how many are there in the class?

16 We each receive $\frac{1}{12}$ of the sweets in a jar. If I get 8 sweets,
how many were in the jar?

17 A glass can hold exactly $\frac{1}{8}$ of a whole carton of juice. If the
amount in the glass is 200 ml, how many litres were in the
carton?

18 A class is divided into 5 equal groups. If there are 4 pupils in a
group, how many pupils are there in the class?

⇨ Finding the original amount (2)

You solved the questions above by using unit fractions.

How would you answer a question such as:

If $\frac{3}{4}$ of an amount is 9, what is the original amount?

Let the original amount be a. Then you can write:

$\frac{3}{4} \times a = 9$

This means that a has been divided by 4 and the result has been multiplied by 3 to get 9

Therefore you must do the opposite. Divide 9 by 3 and then multiply the result by 4 to find a.

$a = 9 \div 3 \times 4$ Do the division first if you can, as you get
$\quad = 12$ smaller numbers.

Check: $12 \div 4 = 3$, $3 \times 3 = 9$

Examples:

(i) If $\frac{2}{5}$ of a number is 8, what is the original number?

$\frac{2}{5}$ of $n = 8$

$\qquad n = 8 \div 2 \times 5$

$\qquad\quad = 20$

Check: $20 \div 5 = 4$, $4 \times 2 = 8$

(ii) If $\frac{3}{8}$ of a bag of sugar weighs 450 g, how many kilograms were there in the original bag?

$\frac{3}{8}$ of $b = 450$

$\qquad b = 450 \div 3 \times 8$

$\qquad\quad = 1200$ g

$\qquad\quad = 1.2$ kg

Check: $1200 \div 8 = 150$, $150 \times 3 = 450$

Exercise 12.6

1 If $\frac{3}{4}$ of an amount is 12, what is the original amount?

2 If $\frac{2}{3}$ of a value is 6, what is the original value?

3 If $\frac{7}{10}$ of a number is 28, what is the original number?

4 If $\frac{5}{12}$ of an amount is 60, what is the original amount?

5 If $\frac{3}{8}$ of a number is 24, what is the original number?

6 A number is divided by 6 and the result is multiplied by 5
The answer is 20. What was the number?

7 A number is divided by 10 and the result is multiplied by 3
The answer is 12. What was the number?

8 A number is multiplied by 5 and the result is divided by 6
The answer is 30. What was the number?

9 A number is multiplied by 7 and the result is divided by 4
The answer is 28. What was the number?

10 A number is divided by 9 and the result is multiplied by 5
The answer is 45. What was the number?

11 We have travelled 135 km and Dad says that we are $\frac{3}{5}$ of the whole way. What is the total length of our journey?

12 $\frac{3}{4}$ of a sum of money is £72
What is the total sum of money?

13 If $\frac{5}{8}$ of a length is 2 m, how long is the whole length?

14 If $\frac{7}{15}$ of a bag of cement weighs 11.2 kg, how heavy is a full bag?

15 If $\frac{5}{12}$ of a volume is 250 ml, how many litres is the total capacity?

16 If $\frac{7}{200}$ of a length is 28 cm, how many metres long is the whole length?

17 I have eaten $\frac{2}{5}$ of a bar of chocolate and only have 180 g left. How many grams were in the whole bar?

18 $\frac{3}{8}$ of the class are away today and there are only 15 of us here. How many of us are in the class when we are all here?

Exercise 12.7: Summary exercise

1 Calculate these amounts.

(a) $\frac{1}{6}$ of 36

(c) $\frac{1}{8}$ of 4 kg, in grams

(b) $\frac{1}{8}$ of 56

(d) $\frac{1}{12}$ of 3 km, in metres

2 Calculate each amount by cancelling. If the answer is not exact write your answer as a mixed number.

(a) $\frac{4}{15}$ of 75

(c) $\frac{3}{7}$ of 56

(b) $\frac{5}{12}$ of 32

(d) $\frac{2}{15}$ of 144

3 Calculate each amount by cancelling. If the answer is not exact write your answer as a decimal.

(a) $\frac{3}{5}$ of 1 kg

(c) $\frac{4}{15}$ of 3 litres

(b) $\frac{5}{8}$ of 2 m

(d) $\frac{5}{12}$ of £9

4 Find the original amount, if:

(a) $\frac{1}{8}$ of the original is 7

(c) $\frac{1}{9}$ of the original is 25p

(b) $\frac{1}{5}$ of the original is 12

(d) $\frac{1}{12}$ of the original is 60 cm.

5 Find the original amount, if:

(a) $\frac{2}{3}$ of the original is 16

(c) $\frac{7}{12}$ of the original is 2.1 kg

(b) $\frac{3}{8}$ of the original is 24

(d) $\frac{8}{15}$ of the original is 4 hours.

6 A 2 kg bag of flour is divided so that $\frac{3}{8}$ is used to make a cake, $\frac{3}{5}$ is used to make a pastry pie crust and the rest is used to decorate the pie. How many grams were used for each recipe?

7 £140 is divided up so that my sister receives $\frac{3}{7}$, my brother receives $\frac{2}{8}$ and I have the rest. How much money do we each get?

8 A number is multiplied by 9 and the result is divided by 5. The answer is 45. What was the number?

9 We planted lots of bean plants but after two weeks $\frac{8}{15}$ of them had died and we only had 105 left. How many did we plant originally?

The Tower of Hanoi puzzle was invented by the French mathematician Edouard Lucas in 1883. If you have a tower of eight discs, of increasing size, stacked on one of three pegs, what is the smallest number of moves you need to move all the discs to the third tower? You can only move one disc at a time and can never put a larger disc on top of a smaller one.

As with many puzzles it is easiest to start with a smaller number and then build up and try to find a pattern. Consider two discs.

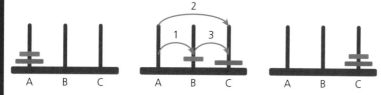

It takes three moves to get the disc from the first tower to the third:

Disc 1 from A to B

Disc 2 from A to C

Disc 1 from B to C

Try this with three discs and then four. Can you predict how many moves will be needed for five discs?

Keep building up your tower until you can find the answer to the minimum number of moves for eight discs. Use diagrams like these to help you.

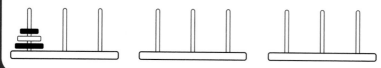

13 Percentages

⇨ Finding the percentage

In Chapter 11, you worked out how to turn a fraction into a percentage. You know that a percentage is a fraction 'out of a hundred'. Using percentages can help you to compare results.

Example:

Last week Josh scored 8 out of 10 for his mental arithmetic test.

This week he scored 9 out of 12. Which was his better result?

To find the answer, rewrite both scores – first as a fraction and then as a percentage.

$$8 \text{ out of } 10 = \frac{8}{10} \qquad\qquad 9 \text{ out of } 12 = \frac{9}{12}$$

$$= 80\% \qquad\qquad\qquad = \frac{3}{4}$$

$$= 75\%$$

Therefore, Josh's score last week was better.

Exercise 13.1

1 Look at this square. It is divided into 100 parts.

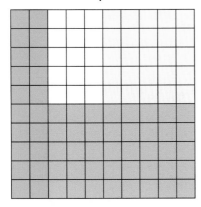

    (a) What percentage is coloured:

        (i) pink

        (ii) yellow

        (iii) blue?

    (b) What percentage is not coloured?

2 Look at this square. It is divided into 50 parts.

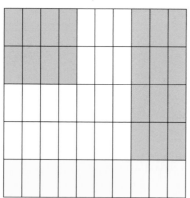

(a) What percentage is coloured:

    (i) pink

    (ii) yellow

    (iii) blue?

(b) What percentage is not coloured?

3 Square A is divided into 25 parts and 7 parts are shaded blue.
Square B is divided into 20 parts and 6 parts are shaded blue.
Calculate the percentage of blue in each square.

4 These are my recent test results. Write each result as a percentage.

English	24 out of 40
Mathematics	63 out of 75
History	36 out of 50
Spelling	16 out of 20

What was my best subject?

5 In a school of 180 pupils, 120 are boys. What percentage of the school are boys?

6 In class 6A, 13 out of 25 pupils are girls. In class 6B, 11 out of 20 pupils are girls.

(a) Which class has the higher percentage of girls?

(b) By how much?

7 140 children were asked what they had for breakfast.
84 had cereal, 28 had eggs, 21 had bananas and the rest had nothing.
Work out what percentage of the children had:

(a) cereal

(b) eggs

(c) bananas

(d) nothing for breakfast.

8 360 children were asked if they liked reading. 288 said 'yes',
54 said 'no' and the rest said they did not know. Work out the
percentages that:

(a) said 'yes'   (c) said 'don't know'.

(b) said 'no'

9 These are the results of a survey on how the pupils at my
school arrive in the morning.

Bus	24
Car	132
Walk	72
Cycle	12

(a) How many pupils took part in the survey?

(b) Write the results as percentages.

10 Our class is learning about healthy eating. We were told that
we need to eat about 2000 calories a day. We looked up the
calories in some of the foods that we enjoy. This is what we
found.

Food	Energy (kcalories)
Packet of crisps	200
Can of cola	250
Glass of orange juice	80
Strawberry yoghurt	160

Calculate what percentage of our daily calories each of the
foods would be.

⇨ Finding a percentage of an amount

In the school prospectus it says that 60% of the pupils take extra
music lessons. If there are 220 pupils in the school, how can you
work out how many take extra music lessons?

The simplest way is to find 10% and then work out 60%.

10% of 220 is 22

60% of 220 is 22 × 6 = 132

For other percentages, it might be easiest to think of them as fractions.

$25\% = \dfrac{1}{4}$ $50\% = \dfrac{1}{2}$ $75\% = \dfrac{3}{4}$

Examples:

(i) Find 50% of 60

$50\% = \dfrac{1}{2}$

$\dfrac{1}{2}$ of 60 = 30

(ii) Find 75% of 120

$75\% = \dfrac{3}{4}$

$\dfrac{1}{4}$ of 120 = 120 ÷ 4

= 30

$\dfrac{3}{4}$ of 120 = 30 × 3

= 90

(iii) Find 40% of 80

10% of 80 = 80 ÷ 10

= 8

40% of 80 = 8 × 4

= 32

(iv) Find 5% of 50

10% of 50 = 50 ÷ 10

= 5

5% of 50 = 5 ÷ 2

$= 2\dfrac{1}{2}$

(v) Find 15% of 24

10% of 24 = 24 ÷ 10

= 2.4

5% of 24 = 2.4 ÷ 2

= 1.2

15% of 24 = 2.4 + 1.2

= 3.6

Exercise 13.2

Calculate these percentages. If the answers are not whole numbers, write them as decimals.

1 10% of 20
2 25% of 16
3 50% of 180
4 75% of 60
5 5% of 40

6 30% of 70
7 80% of 90
8 20% of 120
9 5% of 60
10 90% of 70

11 30% of 15
12 75% of 36
13 15% of 28
14 50% of 25
15 45% of 144

16 40% of 12
17 60% of 25
18 25% of 36
19 65% of 35
20 75% of 144

Percentages as fractions

In the questions above, the percentages were all multiples of 10% or 5%. However, they can be any number.

● Add 12% service charge to a bill of £108

● 24% of the 450 pupils in the school use the school bus.

● There has been a 4% increase in the cost of a school lunch.

You could find any percentage by finding 1% and then multiplying.

Example:

Add 12% to a bill of £108

$$1\% \text{ of } £108 = £1.08$$

$$12\% \text{ of } £108 = £1.08 \times 12$$

$$= £12.96$$

$$\text{Final bill} = £108 + £12.96$$

$$= £120.96$$

It can be easier to write the percentage as a fraction, simplify the fraction to its lowest terms and then calculate.

✱ There are some percentages and fraction equivalents that you know; remember to use them.

Examples:

(i) Find $12\frac{1}{2}\%$ of 72

$$12\frac{1}{2}\% \text{ of } 72 = \frac{1}{8} \text{ of } 72$$

$$= 9$$

(ii) Find $66\frac{2}{3}\%$ of 48

$$66\frac{2}{3}\% \text{ of } 48 = \frac{2}{3} \text{ of } 48$$

$$= 48 \div 3 \times 2$$

$$= 32$$

For other percentages you can write them as a fraction over 100.

Example:

Find 7% of 200

$$7\% \text{ of } 200 = \frac{7}{100_1} \times 200^2$$

$$= 7 \times 2$$

$$= 14$$

Sometimes, you may be able to cancel the percentage with the 100.

Example:

Find 6% of 50

$$6\% \text{ of } 50 = \frac{\overset{3}{6}}{\underset{50_1}{100}} \times 50^1$$

$$= 3$$

In this example, first the percentage is cancelled to become $\frac{3}{50}$ and then the 50s are both divided by 50

You may be left with a remainder. You can write this as a fraction or as a decimal.

 You generally write money and measures as decimals.

Examples:

(i)   Find 9% of 25

$$9\% \text{ of } 25 = \frac{9}{\underset{4}{100}} \times 25^1 \qquad \text{Cancel by 25}$$

$$= \frac{9}{4}$$

$$= 2\frac{1}{4} \text{ or } 2.25$$

(ii)   Find 12% of £15

$$12\% \text{ of } £15 = \frac{\overset{3}{12}}{\underset{25_5}{100}} \times 15^3 \qquad \text{Cancel by 4 and then by 5}$$

$$= \frac{9}{5}$$

$$= £1.80$$

You may find that you have not divided by the largest factor possible at first. That does not matter, you can go back and divide again.

Example:

Find 24% of 450

$$24\% \text{ of } 450 = \frac{24^{6}}{100_{20_{4}}} \times 450^{90^{18}}$$

$$= 6 \times 18$$

$$= 108$$

In this example, both 100 and 450 can be divided by 25, but they have been divided by 5 and then 5 again.

Exercise 13.3

Calculate each amount by first writing the percentage as a fraction. If your answer is not exact then write it as a mixed number.

1 12% of 20

2 14% of 40

3 9% of 180

4 16% of 50

5 $12\frac{1}{2}$% of 80

6 26% of 150

7 72% of 80

8 $33\frac{1}{3}$% of 75

9 16% of 85

10 $37\frac{1}{2}$% of 120

Calculate each amount by first writing the percentage as a fraction. If your answer is not exact then write it as a decimal.

11 8% of £15

12 12% of 450 g

13 15% of 90 cm

14 $66\frac{2}{3}$% of 300 ml

15 9% of £25

16 $62\frac{1}{2}$% of 360 mg

17 18% of £175

18 17% of 225 mm

19 72% of 650 ml

20 16% of 750 g

Changing the units

You may wish to convert the quantities into smaller units. You may be able to divide by 10 or 100 first by cancelling the 0s.

Example:

Find 24% of 3.5 m

$$3.5 \text{ m} = 350 \text{ cm}$$

24% of $350 = \dfrac{24}{100} \times 35\cancel{0}$ First cancel by 10

$$= \dfrac{\cancel{24}^{12}}{\cancel{10}_{5_{1}}} \times \cancel{35}^{7} \quad \text{Cancel by 2 and then by 5}$$

$$= 12 \times 7$$

$$= 84 \text{ cm}$$

Exercise 13.4

Calculate each amount by first writing the percentage as a fraction.

1 $12\frac{1}{2}$% of £6

2 24% of 6 kg

3 8% of 12 m

4 16% of 8 litres

5 6% of 3 tonnes

6 3% of £120

7 $33\frac{1}{3}$% of 4.5 km

8 18% of 3.5 kg

9 19% of 2.4 litres

10 27% of 3 m

11 $2\frac{1}{2}$% of £2

12 95% of 5 litres

13 $66\frac{2}{3}$% of 0.75 kg

14 32% of £6

15 $37\frac{1}{2}$% of 2 km

16 5% of 120 mg

17 12% of £3.25

18 $87\frac{1}{2}$% of 3.2 tonnes

19 $16\frac{2}{3}$% of 0.72 litres

20 $62\frac{1}{5}$% of 0.4 m.

⇨ Percentages and shopping

You will often see signs like this in shop windows. Everyone likes to get a bargain. When a percentage is taken off a price, this is a **discount**.

At other times, you may have to pay a percentage charge, for example, for service in a restaurant.

1 cola	£2.50
1 ginger ale	£2.50
Subtotal:	£5.00
Service charge 20%	£1.00
Total due:	£6.00

Thank you for dining with us!

For **discounts**:

original price	**less**	percentage discount	**equals**	final (sale) price

For **charges**:

original price	**plus**	percentage charge	**equals**	final price

✱ When you calculate **increases**, first calculate the **percentage increase** and then **add** the **increase** to the **original price** of the bill.

Example:

All prices reduced by 10% in today's sale.

If trainers normally cost £36, what will they cost in the sale?

Discount = 10% of £36

$\qquad$ = 36 ÷ 10

$\qquad$ = £3.60

Sale price = £36 − £3.60

$\qquad$ = £32.40

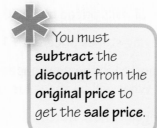

You must **subtract** the **discount** from the **original price** to get the **sale price**.

Exercise 13.5

1 Cereal snack bars have gone up 10%. They used to cost 60p. What is the new price?

2 School jumpers are usually £24 but this week they are being sold with a 20% discount. What is the price this week?

3 On bank holiday Monday the garden centre gives 10% off all prices. What is the cost of:

(a) a plant normally costing £4.50

(b) 12 m of fencing, normally costing £5 a metre

(c) a garden shed, normally costing £225?

4 We went out for a meal and our bill came to £25.40 There was a 10% service charge added on. What was our total bill, including service?

5 There is a 2% charge for paying for a holiday with a credit card. If the cost was £550, what is the final price, with the extra charge?

6 We spent £80 on a meal and left a 15% tip. How much was the bill, including the tip?

7 Value added tax (VAT) is currently charged at 20%. In the builders' merchants, these goods are priced **without** VAT. What will be the price when VAT is added?

(a) A sheet of plywood priced at £17

(b) A box of nails priced at £4.50

(c) 24 paving stones priced at £1.20 each

8 It was reported in the newspaper that house prices are going to fall 15%. What will be the value of these houses after the fall?

You cannot always believe what you read in the newspapers!

(a) A bungalow now worth £150 000

(b) A terraced house now worth £220 000

(c) A large house with a swimming pool, now worth £550 000

9 In the sale, the prices in the computer warehouse have all been cut by 15%. What is the price of:

(a) a laptop that was £399

(b) a printer that was £35

(c) a game that was £12.50?

Give your answers to the nearest whole pound.

⇨ Profit and loss

Before a shop can offer a price at a discount, they must make sure that they can afford it. If they want to make money then they need to make a **profit**. If they sell goods for less than they paid for them, then they will make a **loss** and may find that they owe lots of money.

This is how it works.

	Shopkeeper	
buys from		sells to
Wholesaler or supplier at cost price		Customer at selling price

- To make a **profit**, the selling price must be **more** than the cost price.

- To make a **loss**, the selling price is **less** than the cost price.

Profit and **loss** is often described as a **percentage of the cost price**.

Example:

A shopkeeper buys baked beans for 25p per tin and sells them at a 60% profit. What is the selling price of the baked beans?

Profit = 60% of 25

$= \frac{3}{5} \times 25$ Make it clear at every stage exactly what is being calculated.

= 15p

Selling price = 25p + 15p

= 40p

Some problems ask you to work out the percentage profit or loss. First you must work out the actual profit or loss and then the percentage.

Example:

If a jumper costs £12 from the wholesaler and is sold at £15, what is the percentage profit?

Profit = £15 − £12

= £3

Percentage profit $= \frac{3}{12} \times 100$ You could cancel the fraction to $\frac{1}{4}$

= 300 ÷ 12

= 25%

1 Calculate the selling price of each item.

	Item	Cost price	Profit/loss
(a)	Pack of cards	£1.50	Profit 50%
(b)	Pair of trousers	£15	Loss 10%
(c)	Jumper	£12	Profit $12\frac{1}{2}$%

2 Calculate the percentage profit or loss on each item.

	Item	Cost price	Selling price
(a)	Pen set	£1.50	£2
(b)	Pair of trainers	£16	£24
(c)	Jacket	£35	£28

3 A shopkeeper buys 100 napkins for £10. He sells the napkins in packets of 10 for £2.50 a packet.

(a) What profit does he make on one packet?

(b) What percentage profit does he make on one packet?

4 A corner shop buys 100 kg of potatoes for £50. They sell the potatoes in 5 kg bags at £3.50 a bag.

(a) What profit do they make on the 100 kg of potatoes?

(b) What percentage profit do they make on the 100 kg of potatoes?

5 We are making cakes for the school fair. The ingredients cost us £25 in total. We make 12 cakes and sell each one for £5

(a) What profit do we make overall?

(b) What percentage profit do we make overall?

6 A farmer calculates that keeping enough chickens to produce 100 eggs costs him £5 a week. Every week he sells 100 eggs to the supermarket for £15. The supermarket sells the eggs in packs of 10 for £2.50

   (a) Who makes more profit, the farmer or the supermarket?

   (b) Who makes more percentage profit, the farmer or the supermarket?

7 A shopkeeper buys 12 cans of soup for £9. He sells the soup for £1.20 per can.

   (a) What profit does he make on one can?

   (b) What percentage profit does he make on one can?

8 A garden centre buys a box of 10 seedlings for £4. They sell each seedling at £1.20 each.

   In a frost, 4 of the seedlings died before they could be sold.

   (a) If the garden centre managed to sell the other seedlings, did they make an overall profit or loss?

   (b) What was their profit or loss?

Exercise 13.7: Summary exercise

1 Look at this square. It is divided into 20 parts.

   (a) What percentage is coloured:

      (i)   yellow

      (ii) blue

      (iii) pink?

   (b) What percentage is not coloured?

2 These are Dan's recent test results. Write each result as a percentage.

English   16 out of 40

Mathematics   45 out of 60

French   12 out of 15

3 Calculate these percentages. If the answers are not whole numbers, write them as decimals.

(a) 10% of 25   (c) 30% of 80

(b) 25% of 36   (d) 85% of 70

4 Calculate each amount by first writing the percentage as a fraction. If your answer is not exact then write it as a mixed number.

(a) 16% of 25   (d) 32% of 80

(b) $12\frac{1}{2}$% of 44   (e) $33\frac{1}{3}$% of 108

(c) 28% of 40   (f) 64% of 125

5 Calculate each amount. You may want to write the quantities in smaller units first.

(a) 15% of £3   (d) 8% of 3 litres

(b) 24% of 7 tonnes   (e) 14% of 3 m

(c) $66\frac{2}{3}$% of 9 litres   (f) $37\frac{1}{2}$% of 2 kg

6 There is 20% off the price of everything at the DIY centre today. What is the price of:

(a) an electric drill, normally costing £45

(b) a kitchen cabinet, normally costing £56

(c) a bathroom suite, normally costing £240?

7 A restaurant adds $12\frac{1}{2}$% service charge to all bills over £30 What is the total price for:

(a) a meal costing £120 without service

(b) four set meals at £15 each

(c) a main course at £15 and a drink costing £4.50?

8 A greengrocer buys a box of 12 cabbages for £9
   He sells them for £1.20 each.

   (a) What percentage profit does he make on one cabbage?

   (b) The greengrocer did not manage to sell the last 5 cabbages
       before they went yellow. What was his overall percentage
       profit or loss?

Activity – Let's get into shapes!

Cube

Draw this net for a cube on square spotted paper. Add the tabs so
you can fold it up into a solid.

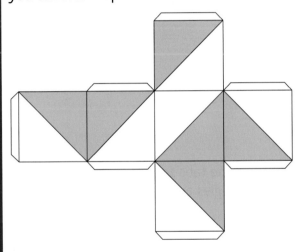

You can decorate yours differently but do make sure it is
geometric and interesting.

Cube investigation

There are 11 different nets of a cube altogether. Try to work out
the other 10

Tetrahedron

Draw this net onto triangular spotted paper. Include the tabs so you can fold it up into a solid. Decorate your net carefully before you fold it up.

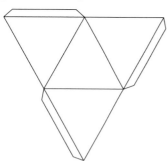

Tetrahedron investigation

There are only two possible nets of a tetrahedron, one is shown above. Which of these will also make a tetrahedron?

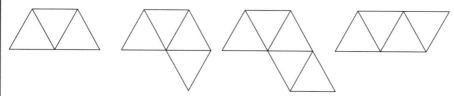

Octahedron

Copy this net onto triangular spotted paper. Include the tabs so you can fold it up into a solid. Make sure all your triangles are equilateral and decorate them first.

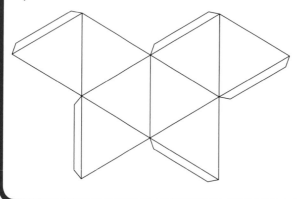

 # Negative numbers

⇨ Which way is up?

What do you know about mountains?

The world's highest mountains are:

● **Mount Everest**, with a height of 8848 m

● **K2**, with a height of 8611 m

● **Kangchenjunga**, with a height of 8586 m.

But where are these heights measured from?

The heights are given as **metres above sea level**, but where is sea level?

Sea level is the level of the seas and oceans, as an average between high and low tides. It is calculated by measuring the level of the ocean over extended periods of time and under all types of weather conditions.

This is a chart of the Yantlet Channel, at the entrance to the River Thames.

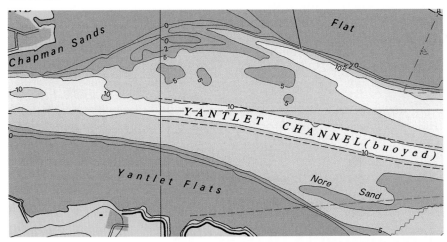

Disclaimer: Not to be used for navigation

The line between the blue and the green areas is taken as sea level.

Sea level is assigned the value 0

Heights above sea level are positive and the depths below sea level are negative.

The section through the Yantlet Channel shows this.

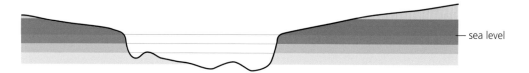

— sea level

Exercise 14.1

Use two dice for this exercise. Mark one with negative numbers from ⁻1 to ⁻6 and the other with positive numbers from 1 to 6

1 Draw a vertical line down the side of your exercise book. Mark it from ⁻10 up to ⁺10

2 Draw a line one square wide at zero.

3 Throw your two dice. Draw a line **up** the number of squares shown on the positive die. Then draw a horizontal line one square wide. Now draw a line **down** the number of squares on the negative die.

   If you had thrown ⁺3 and ⁻5, your lines would look like this.

   ⁺3 + ⁻5 = ⁻2

4 Continue throwing the dice and marking your positions, according to your scores, until you reach the end of the page.

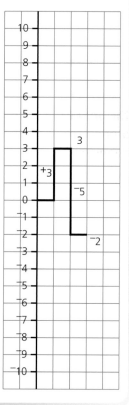

⇨ Adding positive numbers to negative numbers

To add positive numbers on a number line, count from zero (0) to the first number, then count forward by the second number. The answer is the point where you end up.

This works just as well when starting with a negative number.

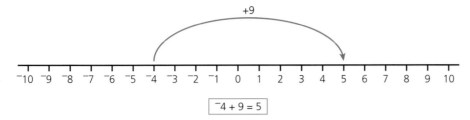

$$^-4 + 9 = 5$$

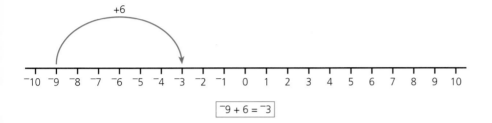

$$^-9 + 6 = ^-3$$

Exercise 14.2

Calculate the answer to each addition. First draw a number line across your page and use that for the first few questions, then do the calculations without using a number line.

1 $^-1 + 5$

2 $^-1 + 8$

3 $^-9 + 3$

4 $^-6 + 4$

5 $^-6 + 3$

6 $^-5 + 8$

7 $^-7 + 9$

8 $^-4 + 8$

9 $^-2 + 6$

10 $^-7 + 10$

11 $^-11 + 2$

12 $^-13 + 4$

13 $^-19 + 13$

14 $^-21 + 25$

15 $^-36 + 5$

16 $^-15 + 18$

17 $^-27 + 9$

18 $^-24 + 28$

19 $^-29 + 16$

20 $^-35 + 40$

⇨ Subtraction with negative numbers

To subtract a positive number on a number line, count from zero (0) to the first number, then count back by the second number. The answer is the point where you end up. This works just as well with negative numbers.

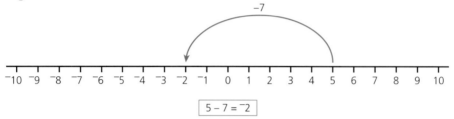

$$5 - 7 = {}^-2$$

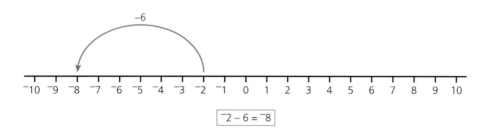

$${}^-2 - 6 = {}^-8$$

Exercise 14.3

Calculate the answer to each subtraction. First draw a number line across your page and use that for the first few questions, then do the calculations without using a number line.

1 7 − 5	**6** 4 − 9
2 1 − 5	**7** 7 − 9
3 ⁻1 − 5	**8** ⁻4 − 5
4 6 − 7	**9** ⁻1 − 6
5 ⁻2 − 3	**10** 4 − 11

11 ⁻11 − 2	**16** ⁻15 − 24
12 ⁻15 − 4	**17** ⁻34 − 8
13 ⁻23 − 11	**18** 13 − 28
14 15 − 20	**19** 25 − 30
15 14 − 23	**20** ⁻35 − 43

⇨ Adding and subtracting negative numbers

In the previous exercises, you added or subtracted positive numbers. What happens if you have to add or subtract a negative number?

On the number line we face right and move forward when we are adding and face left and move forward when we are subtracting.

Think what happens on a number line. Imagine yourself as one of these people. To add a positive number, you will face towards the right and move forward. To subtract a positive number, you will face towards the left and move forward.

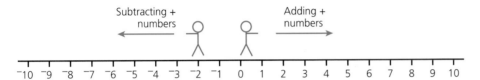

Now, if you face **right** and walk **forwards** to add a **positive** number, you must still face right, but to add a **negative** number you face **right** and walk **backwards**.

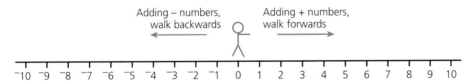

Therefore adding a negative number is the same as subtracting the 'opposite' positive number.

> ## Examples:
>
> (i) Add: $10 + (^-4)$
>
> $10 + (^-4) = 10 - 4$
>
> $= 6$
>
> (ii) Add: $^-3 + (^-5)$
>
> $^-3 + (^-5) = ^-3 - 5$
>
> $= ^-8$

If you face left and walk forwards to subtract a positive number, you must face left and walk backwards to subtract a negative number.

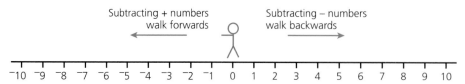

Subtracting + numbers
walk forwards

Subtracting – numbers
walk backwards

Therefore subtracting a negative number is the same as adding the 'opposite' positive number.

Examples:

(i) Subtract: $7 - (^-2)$

$$7 - (^-2) = 7 + 2$$

$$= 9$$

(ii) Subtract: $^-8 - (^-5)$

$$^-8 - (^-5) = ^-8 + 5$$

$$= ^-3$$

This might seem strange but think about double negatives.

● 'I will not miss lunch today,' means that I will have lunch.

● 'I won't not be your friend,' means that you will be a friend.

A double negative is a positive!

Exercise 14.4

Calculate the answers. You may find a number line helpful for the first few questions.

1 $7 + (^-5)$ 6 $^-5 - (^-5)$

2 $^-1 + (^-2)$ 7 $3 + (^-7)$

3 $4 - (^-3)$ 8 $^-6 + (^-5)$

4 $^-3 - (^-4)$ 9 $^-8 - (^-3)$

5 $^-6 + (^-3)$ 10 $7 + (^-1)$

11 $^-15 + (^-20)$	16 $17 + (^-32)$
12 $^-18 - (^-12)$	17 $^-19 - (^-12)$
13 $^-25 + (^-10)$	18 $28 + (^-23)$
14 $35 + (^-13)$	19 $^-25 - (^-27)$
15 $^-43 - (^-32)$	20 $^-16 - (^-18)$
21 $^-5 - (^-3) + (^-1)$	26 $12 - (^-14) + (^-15)$
22 $4 + (^-3) - (^-7)$	27 $24 - (^-19) + (^-15)$
23 $^-10 + (^-4) - (^-15)$	28 $17 - (^-15) - (^-27)$
24 $^-6 - 4 - (^-3)$	29 $^-19 + (^-17) - (^-15)$
25 $^-3 + (^-4) - 7$	30 $23 + (^-19) - (^-16)$

⇨ Temperatures

Negative numbers are used to describe temperatures that drop below zero. Look at the temperatures you are given and use them to answer the questions. Remember to write down your calculations carefully. You can use brackets so that you do not get the signs confused.

Example:

The temperature is $^-15$ °C in Moscow and $^-12$ °C in Leningrad.

What is the difference in the temperatures?

Difference = $^-12 - (^-15)$

= $^-12 + 15$

= 3

So the difference is 3 degrees.

Exercise 14.5

Look at this chart, then use it to answer questions 1–6

■ Lowest recorded temperatures in some European countries

	Austria	⁻52 °C
	France	⁻41 °C
	Germany	⁻37 °C
	Greece	⁻27 °C
	Malta	2 °C

1 Which of the countries had the lowest temperature ever recorded?

2 Which of the lowest temperatures recorded in the table was the highest?

3 How much lower was the temperature in Austria than in Malta?

4 What is the difference between the temperatures in Greece and in Germany?

5 How much higher was the lowest temperature in Malta than the lowest temperature France?

6 The lowest temperature recorded in Russia was 6 degrees lower than that in Austria. What was the lowest temperature recorded in Russia?

Look at this chart, then use it to answer questions 7–11

■ Highest temperatures in February in East Coast USA

Maine	⁻12 °C
Connecticut	⁻5 °C
New York	⁻8 °C
West Virginia	5 °C
Georgia	11 °C
Florida	15 °C

VERMONT
MAINE
NEW HAMPSHIRE
MASSACHUSETTS
WISCONSIN
MICHIGAN
NEW YORK
RHODE ISLAND
CONNECTICUT
PENNSYLVANIA
NEW JERSEY
OHIO
DELAWARE
ILLINOIS INDIANA
WEST
VIRGINIA
MARYLAND
VIRGINIA
KENTUCKY
Washington, DC
NORTH CAROLINA
TENNESSEE
SOUTH
CAROLINA
MISSISSIPPI
GEORGIA
ALABAMA
LOUISIANA
FLORIDA

N

7 Which of the states had the lowest temperature in February?

8 How much warmer was it in Florida than in Maine?

9 What was the difference in temperature between New York and Connecticut?

10 What was the difference in temperature between Georgia and Connecticut?

11 The temperature in Michigan was 13 degrees lower than that in Maine. What was the temperature in Michigan?

⇨ Number patterns

Look at this number line. It goes below 0 and runs from ⁻1 to 3 in steps of 0.4

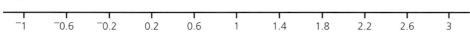

| ⁻1 | ⁻0.6 | ⁻0.2 | 0.2 | 0.6 | 1 | 1.4 | 1.8 | 2.2 | 2.6 | 3 |

Number patterns sometimes include decimal fractions.

Sometimes they have terms that are below 0

- 0, 0.3, 0.6, 0.9 What would come next? 0.9 + 0.3 = 1.2

- ⁻2, ⁻1.6, ⁻1.2, ⁻0.8 What would come next? ⁻0.8 + 0.4 = ⁻0.4

To complete number patterns you need to:

1 Work out what the step is between two adjacent terms in the pattern.

2 Add or subtract to find the missing term.

Examples:

Find the missing terms in each pattern.

(i) 2.7, 3, 3.3, 3.6, ..., ...

    The step is $3.6 - 3.3 = 0.3$

    The missing terms are: $3.6 + 0.3 = 3.9$

    and: $3.9 + 0.3 = 4.2$

    The pattern is: 2.7, 3, 3.3, 3.6, **3.9, 4.2**

(ii) ⁻1.5, ..., ⁻0.7, ⁻0.3, ..., 0.5

    The step is the difference between ⁻0.7 and ⁻0.3 so it is 0.4

    The missing terms are: $0.5 - 0.4 = 0.1$

    And: $⁻0.7 - 0.4 = ⁻1.1$

    The pattern is: ⁻1.5, **⁻1.1**, ⁻0.7, ⁻0.3, **0.1**, 0.5

Exercise 14.6

Find the missing terms in each pattern.

1 ⁻8, ⁻4, 0, 4, ..., ...

2 ⁻15, ⁻12, ⁻9, ⁻6, ..., ...

3 ..., ..., ⁻4, ⁻1, 2, 5

4 ..., ⁻17, ⁻11, ⁻5, 1, ...

5 ..., ⁻25, ..., ⁻11, ⁻4, 3

6 ⁻3.5, ⁻3, ⁻2.5, ⁻2, ..., ...

7 ⁻2.4, ⁻2.0, ⁻1.6, ⁻1.2, ..., ...

8 ⁻2.7, ⁻1.9, ⁻1.1, ⁻0.3, ..., ...

9 ..., ..., 0, 0.5, 1, 1.5

10 ..., ..., 0, 0.7, 1.4, 2.1

11 ..., ..., 0.3, 0.8, 1.3, 1.8

12 ..., ..., ⁻0.4, 0.5, 1.4, 2.3

13 ..., ⁻0.6, 0, ..., 1.2, 1.8

14 ⁻1.2, ..., 0.2, 0.9, 1.6, ...

15 ⁻1.6, ..., 0.8, ..., 3.2, 4.4

16 ..., 1, 2.5, 4, 5.5, ...

17 ..., 0, ..., 1.4, 2.1, 2.8

18 ..., 1, 2.2 ..., 4.6, 5.8

19 ⁻2.3 ..., ⁻0.7, ..., 0.9, 1.7

20 ⁻3.5, ..., ⁻1.1, 0.1, ..., 2.5

Exercise 14.7: Summary exercise

Copy the questions and complete the calculations.

1 7 – 9

2 ⁻3 + 6

3 ⁻7 + 3

4 5 – (⁻7)

5 ⁻3 + (⁻8)

6 ⁻11 – 16

7 8 – 25

8 ⁻29 + (⁻27)

9 ⁻37 – (⁻19)

10 19 – (⁻17)

Look at this chart, then use it to answer questions 11–14

■ Temperatures in some European cities in January

Moscow	⁻11 °C
Oslo	⁻5 °C
London	⁻1 °C
Paris	5 °C
Athens	17 °C

11 How much warmer is it in Athens than Moscow?

12 What is the difference in temperature between London and Oslo?

13 It is 15 degrees colder on Mont Blanc than in Paris. What is the temperature on Mont Blanc?

14 The temperature at Archangel is 13 degrees lower than the temperature in Moscow. What is the temperature in Archangel?

15 Find the missing terms in each pattern.

(a) ⁻10, ⁻7, ⁻4, ..., ..., 5

(b) ⁻2.4, ⁻1.8, ⁻1.2, ⁻0.6, ...

(c) ⁻2.4, ⁻1.9, ⁻1.4, ⁻0.9, ..., ...

(d) ..., ..., ⁻0.3, 0.1, 0.5, 0.9

Activity – Rangoli patterns

Hindu and Sikh families often decorate their homes with Rangoli patterns during festivals such as Divali. Some patterns are pictorial whereas others are based on geometrical patterns.

For celebrations the patterns may be made on the door step from ground rice, lentils, split peas, seeds and perfumed red powder. Then everyone coming into the house would pass over the pattern and – it is believed – have good luck.

After a design has been drawn on dotted paper it can be used as a basis for collage work, block printing, embroidery or batik work.

Making your own Rangoli patterns

1 On square spotted paper, draw two lines to divide your paper into four quarters. These are to be lines of symmetry, one horizontal and one vertical.

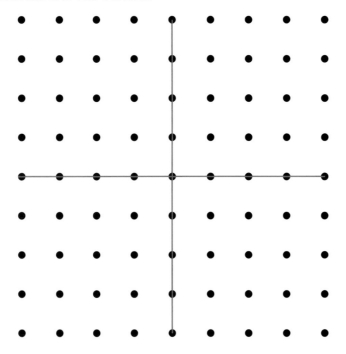

2 Join the dots in one quarter of your paper together to make an interesting shape – do not make it too complicated! You might want to start with just two lines first.

3 Now reflect the shape you have drawn in the vertical line of symmetry.

4 Next reflect **both** patterns in the horizontal line of symmetry.

Steps 1–2

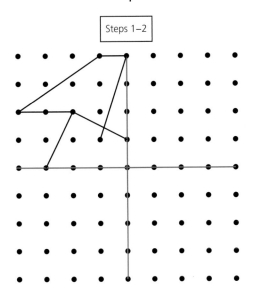

Steps 3–4

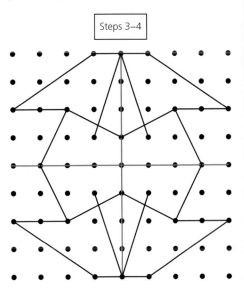

5 Rub out the lines of symmetry and draw in the diagonals. These are the new lines of symmetry. Complete your pattern by reflecting in these lines.

6 When your pattern is complete rub out the diagonals and then carefully colour your final design so that the regions are clear.

Step 5

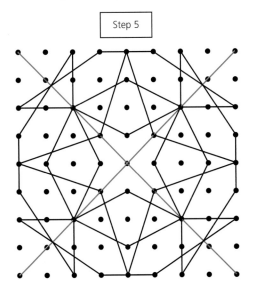

Step 6

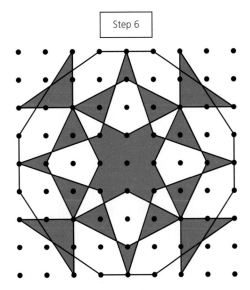

15 Transformations

⇨ What are transformations?

Transformation is a general term for a movement of a two-dimensional **object** to produce a resulting **image**. You can create attractive artwork by doing this. Mathematically you can do this on a co-ordinate grid.

⇨ Negative co-ordinates

This co-ordinate grid has horizontal and vertical axes marked with values from ⁻6 to 6

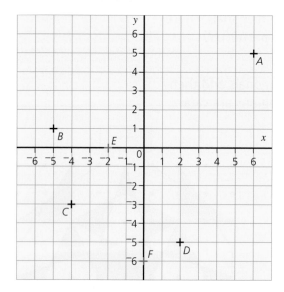

Points *A*, *B*, *C* and *D* are marked. You can write these points as *A*(6, 5), *B*(⁻5, 1), *C*(⁻4, ⁻3) and *D*(2, ⁻5). The numbers in the brackets tell you the co-ordinates of the points.

Now look at points *E* and *F*, which are on the axes.

E is the point (⁻2, 0) on the *x*-axis, where the value of *y* is 0

F is the point (0, ⁻6) on the *y*-axis, where the value of *x* is 0

Exercise 15.1

1 Look at this co-ordinate grid. Write down the co-ordinates of points A–K.

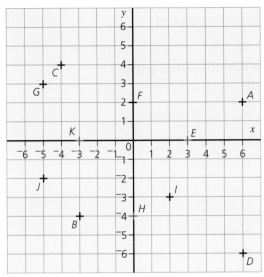

For each of questions 2–6, you will need a blank copy of the co-ordinate grid above.

2 On your co-ordinate grid, mark these points.

$A(3, 5)$ $B(^-4, 1)$ $C(^-2, ^-4)$ $D(6, ^-2)$ $E(^-5, 0)$

$F(0, 3)$ $G(^-3, 5)$ $H(5, 0)$ $I(3, ^-4)$ $J(0, ^-2)$

3 **(a)** On a new co-ordinate grid, mark the points:
   $A(5, 3), B(^-1, 3), C(^-1, ^-2)$

   (b) Join A to B and B to C.

   (c) Mark a point D so that $ABCD$ is a rectangle.

   (d) Write down the co-ordinates of D.

4 **(a)** On a new co-ordinate grid, mark the points:
   $A(1, ^-3), B(^-4, ^-3), C(^-4, 2)$

   (b) Join A to B and B to C.

   (c) Mark a point D so that $ABCD$ is a square.

   (d) Write down the co-ordinates of D.

⇨ Reflection

A **reflection** is the **image** of an **object** viewed in a mirror. The resulting image is **identical** to the original object except that it is **reversed**.

Mathematically, reflection is the image of an object in a **line of symmetry**.

On this grid, triangle X is a reflection of triangle A in the x-axis and triangle Y is a reflection of triangle A in the y-axis.

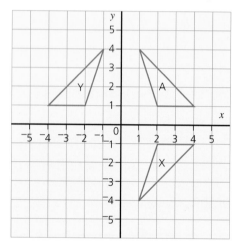

Exercise 15.2

1 (a) Copy this grid and the triangles into your exercise book.

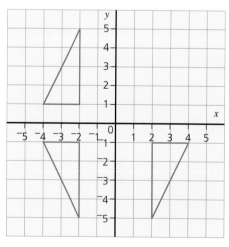

✳ You can use a mirror to check your reflections in this exercise.

(b) Triangle X is a reflection of triangle A in the x-axis and triangle Y is a reflection of triangle A in the y-axis. Label the triangles A, X and Y.

2 (a) Copy this grid and triangle A into your exercise book.

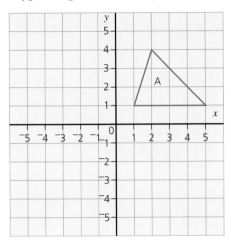

(b) Draw a reflection of the triangle in the x-axis. Label it X.

(c) Draw a reflection of the triangle in the y-axis. Label it Y.

For each of questions 3–6, you will need a new, blank co-ordinate grid, like this.

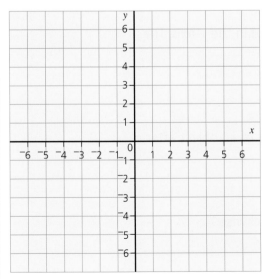

3 (a) Draw quadrilateral C with vertices at (1, 1), (2, 4), (3, 3) and (4, 1)

(b) Draw a reflection of the quadrilateral in the x-axis. Label it X.

(c) Draw a reflection of the quadrilateral in the y-axis. Label it Y.

4 (a) Draw shape D with vertices at (⁻5, ⁻1), (⁻5, ⁻3), (⁻3, ⁻5), (⁻1, ⁻3) and (⁻1, ⁻1)

(b) Draw a reflection of the shape in the x-axis. Label it X.

(c) Draw a reflection of the shape in the y-axis. Label it Y.

5 Draw three tables in your exercise book. Use them to record the co-ordinates of the vertices in questions 2–4. This table for question 2 has been started for you.

Triangle A	Triangle X	Triangle Y
(1, 1)	(1, ⁻1)	
(2, 4)		(⁻2, 4)
(5, 1)		

6 Copy and complete these sentences.

An object with co-ordinates (a, b) is reflected in the x-axis. The co-ordinates of the image are: (... , ...)

An object with co-ordinates (a, b) is reflected in the y-axis. The co-ordinates of the image are: (... , ...)

⇨ Rotation

A rotation is a turn. To be able to draw an image of an object after a rotation, you need to know:

- the centre of rotation
- the angle of rotation
- the direction of rotation – clockwise or anticlockwise.

Clockwise

Anticlockwise

Look at the shapes on this co-ordinate grid. Use tracing paper to check these statements.

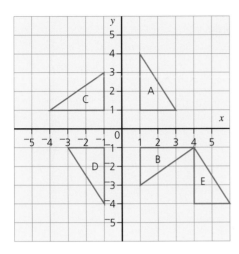

B is the image of A after a rotation of 90° clockwise about the origin.

C is the image of A after a rotation of 90° anticlockwise about the origin.

D is the image of A after a rotation of 180° about the origin.

E is the image of B after a rotation of 90° anticlockwise about (4, ⁻1)

* For a rotation of 180° it does not matter if the direction is clockwise or anticlockwise.

Exercise 15.3

1 Look at the images on this grid.

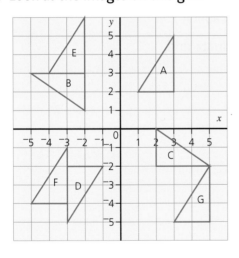

* You can use tracing paper to check your rotations in this exercise.

Write in full the rotation that maps:

(a) triangle A to triangle B

(b) triangle B to triangle C

(c) triangle A to triangle D

(d) triangle B to triangle E

(e) triangle F to triangle D

(f) triangle G to triangle C

(g) triangle C to triangle G

2 (a) Copy this grid and object A into your exercise book.

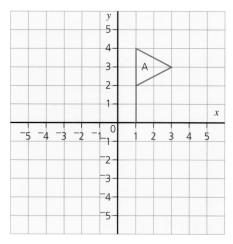

(b) Draw the image of A after a rotation of 90° clockwise about the origin. Label it B.

(c) Draw the image of A after a rotation of 90° anticlockwise about the origin. Label it C.

(d) Draw the image of A after a rotation of 180° about the origin. Label it D.

(e) Draw a reflection of the object in the y-axis. Label it Y.

For questions 3 and 4, you will need a blank co-ordinate grid, like this.

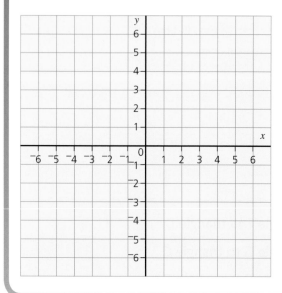

3 (a) Draw triangle A with vertices at (⁻4, 1), (⁻1, 1) and (⁻4, 4)

(b) Draw the image of A after a rotation of 90° clockwise about the origin. Label it B.

(c) Draw the image of A after a rotation of 90° anticlockwise about the origin. Label it C.

(d) Draw the image of A after a rotation of 180° about the origin. Label it D.

4 (a) Draw quadrilateral A with vertices at (1, 1), (1, 2), (4, 2) and (4, 1)

(b) Draw the image of A after a rotation of 90° clockwise about the origin. Label it B.

(c) Draw the image of A after a rotation of 90° anticlockwise about the origin. Label it C.

(d) Draw the image of A after a rotation of 180° about the origin. Label it D.

(e) Draw the image of A after a rotation of 90° clockwise about (4, 2). Label it E.

(f) Draw the image of B after a rotation of 180° about (2, ⁻3) Label it F.

5 Draw three tables in your exercise book. Use them to record the co-ordinates of the vertices in questions 2–4. This table for question 2 has been started for you.

Object A	Object B 90° c/w	Object C 90° a/c	Object D 180°
(1, 0)	(⁻1, 0)		
(1, 2)			
(1, 4)			
(3, 3)			

6 Copy and complete these sentences.

An object with co-ordinates (a, b) is rotated 90° clockwise about the origin. The co-ordinates of the image are: (... , ...)

An object with co-ordinates (a, b) is rotated 90° anticlockwise about the origin. The co-ordinates of the image are: (... , ...)

An object with co-ordinates (a, b) is rotated 180° anticlockwise about the origin. The co-ordinates of the image are: (... , ...)

7 Discuss with your partner what would happen if you kept on rotating a shape past 180°. Would it ever come back to the position it started from? Could you describe a rotation of 90° anticlockwise as a clockwise rotation? How many degrees clockwise would the image need to turn?

⇨ Translations

A **translation** is a movement. You can describe the movement of a point or a shape:

● first by moving forward or back horizontally (in the direction of the *x*-axis)

● then by moving up or down vertically (in the direction of the *y*-axis).

Look at the shapes on this grid.

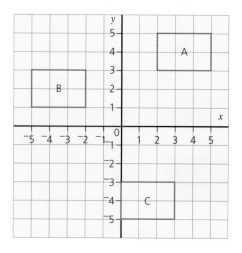

Rectangle A maps to rectangle B by a translation of 7 squares to the left and 2 down.

Rectangle B maps to rectangle C by a translation of 5 squares to the right and 6 down.

Rectangle C maps to rectangle A by a translation of 2 squares to the right and 8 up.

Exercise 15.4

1 Look at the images on this grid.

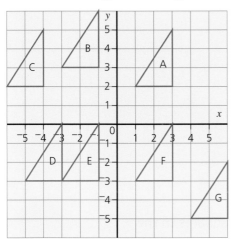

Write in full the translation that maps:

(a) triangle A to triangle B

(b) triangle A to triangle C

(c) triangle A to triangle D

(d) triangle A to triangle F

(e) triangle A to triangle G

(f) triangle A to triangle E

(g) triangle D to triangle E

(h) triangle F to triangle A

(i) triangle G to triangle B

(j) triangle E to triangle B.

2 (a) Copy this grid and object A into your exercise book.

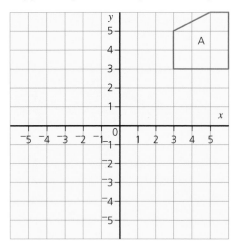

(b) Draw the image of A after a translation of 7 squares to the left and 2 down. Label it B.

(c) Draw the image of A after a translation of 4 squares to the left and 5 down. Label it C.

(d) Draw the image of A after a translation of 8 squares to the left and 8 down. Label it D.

For each of questions 3–5, you will need a new, blank co-ordinate grid, like this.

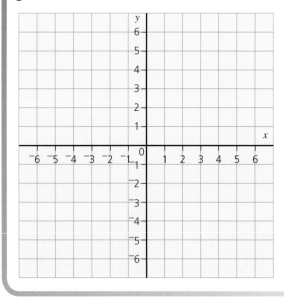

3 (a) Draw triangle A with vertices at ($^-$4, 1), ($^-$1, 1) and ($^-$3, 4)

   (b) Draw the image of A after a translation of 6 squares to the right and 2 up. Label it B.

   (c) Draw the image of A after a translation of 1 square to the left and 3 down. Label it C.

   (d) Draw the image of A after a translation of 5 squares to the right and 5 down. Label it D.

   (e) Draw the image of D after a translation of 5 squares left and 1 down. Label it E.

   (f) Describe the translation that maps A to E.

4 (a) Draw triangle A with vertices at (5, $^-$2), (1, $^-$1) and (2, $^-$4)

   (b) Draw the image of A after a translation of 5 squares to the left and 5 up. Label it B.

   (c) Draw the image of A after a translation of 4 squares to the left. Label it C.

   (d) Draw the image of A after a translation of 6 squares up. Label it D.

   (e) Draw the image of D after a translation of 7 squares left and 8 down. Label it E.

   (f) Describe the translation that maps A to E.

5 (a) Draw quadrilateral A with vertices at (1, 1), (1, 3), (4, 3) and (4, 1)

   (b) Draw the image of A after a translation of 4 squares to the left. Label it B.

   (c) Draw the image of A after a translation of 4 squares down. Label it C.

   (d) Draw the image of A after a translation of 5 squares to the left and 5 down. Label it D.

   (e) Draw the image of D after a translation of 1 square to the left and 7 up. Label it E.

   (f) Describe the translation that maps A to E.

6 Draw a table like this in your exercise book. Use it to record the co-ordinates of the vertices in questions 3–4

Object A	Move right (+) / left (−)	Move up (+) / down (−)	Images
(⁻4, 1)	+ 6 − 1 + 5	+ 2 − 3 − 5	B (2, 3) C (⁻5, ⁻2) D (1, ⁻4)
(⁻1, 1)			B C D
(⁻3, 4)			B C D

7 (a) Discuss your results with your partner. Write down some rules to explain how you can check that your image has been correctly drawn after a single translation.

(b) Copy and complete these sentences.

An object with co-ordinates (a, b) is translated c squares to the right and d squares up. The co-ordinates of the image are: (... , ...)

An object with co-ordinates (a, b) is translated c squares to the left and d squares up. The co-ordinates of the image are: (... , ...)

An object with co-ordinates (a, b) is translated c squares to the right and d squares down. The co-ordinates of the image are: (... , ...)

An object with co-ordinates (a, b) is translated c squares to the left and d squares down. The co-ordinates of the image are: (... , ...)

Exercise 15.5: Summary exercise

1 Look at the images on this grid.

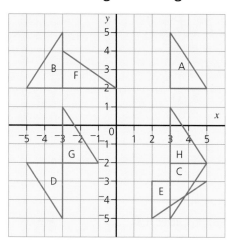

A **transformation** can be a reflection, a rotation or a translation.

(a) What reflection maps object A to image B?

(b) What reflection maps object A to image C?

(c) What rotation maps object A to image D?

(d) What rotation maps object A to image E?

(e) What rotation maps object B to image F?

(f) Write in full the translation that maps object A to image G.

(g) Write in full the translation that maps object A to image H.

(h) What rotation maps object D to image G?

2 (a) Copy this grid and object A into your exercise book.

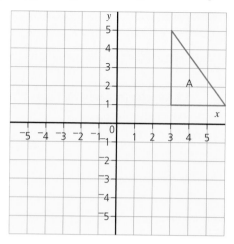

(b) Draw the image of A after a reflection in the *y*-axis. Label it B.

(c) Draw the image of A after a rotation of 90° clockwise about the origin. Label it C.

(d) Draw the image of A after a translation of 7 squares to the left and 4 down. Label it D.

(e) Draw the image of D after a translation of 1 square left and 3 down. Label it E.

(f) Describe the translation that maps A to E.

For each of questions 3 and 4 copy the grid from question 2 into your exercise book.

3 (a) Draw shape A with vertices at (⁻5, ⁻1), (⁻5, ⁻3), (⁻3, ⁻5), (⁻1, ⁻3) and (⁻1, ⁻1)

(b) Draw the image of A after a reflection in the *x*-axis. Label it B.

(c) Draw the image of A after a rotation of 180° about the origin. Label it C.

(d) Draw the image of A after a translation of 6 squares to the right. Label it D.

(e) What transformation maps object B to image D?

4 (a) Draw shape A with vertices at ($^-$2, 0) ($^-$3, 2) ($^-$2, 4) and ($^-$1, 2)

(b) Draw the image of A after a reflection in the y-axis. Label it B.

(c) Draw the image of A after a rotation of 180° about the origin. Label it C.

(d) Draw the image of A after a translation of 6 squares to the right and 4 down. Label it D.

(e) What rotation maps object C to image D?

Activity – Tessellations

The artist M.C. Escher is famous for his 'impossible constructions' – drawings of strange buildings with flights of stairs hanging upside-down or joined in loops. He is also famous for arranging shapes of animals in interesting patterns.

You can see many examples of Escher's work on the internet.

Here is a simple imitation of Escher's pattern style:

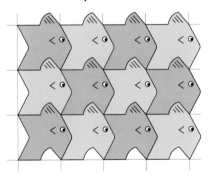

Patterns like these are called tessellations, from the Latin word *tesserae*, meaning a small square tile.

Squares are the simplest and most common form of tessellation.

This is how to build up a tessellation from a square.

You can use squared paper, or use a graphics package on a computer.

Draw an outline on one side of a square.

Draw an exact copy on the opposite side of the same square.

Draw a different outline on the base of the square.

Draw an exact copy on the top of the square.

Now copy your shape, to form a tessellation.   Decorate your tessellation.

 # 16 Formulae

⇨ What is a formula?

In some of the exercises you have worked through you have been asked to write a rule, in words.

For example, in the last chapter you wrote some rules for translations.

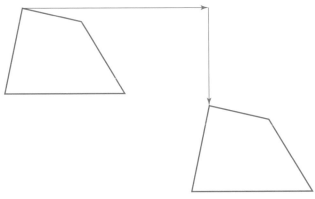

- Add the number of squares that you move horizontally to the x-co-ordinate to get the x-co-ordinate of the image.

- Add the number of squares that you move vertically to the y-co-ordinate to get the y-co-ordinate of the image.

Each of those takes quite a lot of words.

It is often possible to write instructions more simply, using letters, in a **formula**.

If the original co-ordinates are (x, y) and you move a squares right and b up, then the co-ordinates of the image will be $(x + a, y + b)$.

You will often find it helpful to use a formula to solve a problem.

 The plural of formula is **formulae**.

 In textbooks, when letters represent unknown numbers, they are written in italics.

Examples:

How many days are there in:

  (i) 1 week   (ii) 5 weeks   (iii) n weeks?

  (i) There are 7 days in 1 week.

  (ii) There are 7×5 days = 35 days in 5 weeks.

 (iii) There are $7 \times n$ days in n weeks.

Exercise 16.1

1 (a) How many pence are there in £1?

  (b) How many pence are there in: (i) £5   (ii) £7?

  (c) Write a formula for the number of pence in £n.

2 (a) How many grams are there in 1 kg?

  (b) How many grams are there in: (i) 5 kg   (ii) 9 kg?

  (c) Write a formula for the number of grams in m kg.

3 (a) If 12 is one dozen, how much is: (i) 2 dozen (ii) 5 dozen?

  (b) Write a formula for the number in d dozen.

4 (a) How many days are there in 1 week?

  (b) How many weeks are there in: (i) 56 days (ii) 140 days?

  (c) Write a formula to work out the number of weeks in d days.

5 (a) A car travels at 30 kph. How long does it take to travel:
    (i) 10 km   (ii) 60 km?

  (b) Write a formula for the time the car takes to travel d km.

  (c) Write a formula for the time a car travelling at s kph takes
     to travel a distance d kilometres.

6 I am 5 years older than my brother. My brother is x years old.
  Write an expression in terms of x for my age.

7 My sister is 3 years younger than I am. I am y years old.
Write a formula for my sister's age.

8 I am y years older than my brother. My brother is x years old.
Write a formula for my age.

9 My sister is b years younger than I am. I am y years old.
Write a formula for my sister's age.

10 The sum of my parents' ages is 64. My father is a years old.
Write a formula for my mother's age.

⇨ Using brackets

Look at this formula.

$3 \times n + 2$

It is not clear if this means multiply 3 by n and then add 2 or add 2 to n and multiply the result by 3. Following the rules of BIDMAS, you would multiply before you add. But what if you wanted to add before you multiply?

Using brackets makes it clear.

When writing formulae, you can leave out the $\times$ sign. Instead of $3 \times n$, just write $3n$. Similarly, instead of using $\div$ signs, write expressions involving division as fractions.

You would write:

● 'multiply 3 by n and then add 2' as $3n + 2$

● 'add 2 to n and multiply the result by 3' as $3(n + 2)$

● 'add 2 to a and divide by b' as $\frac{a+2}{b}$

Write these word sentences as formulae. Use brackets when appropriate.

1 Multiply 2 by n.

2 Add 4 to m.

3 Divide p by 6

4 Take 5 away from a.

5 Add 4 to m and multiply the result by 5

6 Take 5 away from a and multiply the result by 2

7 Multiply n by 2 and add 5

8 Take b away from a and divide the result by c.

9 Add n to m and multiply the result by 5

10 Add s to t and multiply the result by n.

11 Multiply 2 by n and add m to the result

12 Add 3 to m and divide the result by p.

13 Divide p by 5 and multiply the result by 4

14 Take 6 away from b and multiply the result by a.

15 Divide a by b and add c to the result.

16 Take m away from n and multiply the result by p.

⇨ Formulae and patterns

When you are working with patterns you can sometimes spot a general formula. For example here are some dots arranged in squares.

The next pattern will have 6 × 6 dots and the nth pattern would have $n \times n$ dots.

The total number of dots $d = n \times n$ or $d = n^2$

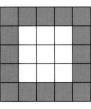

Exercise 16.3

1 (a) Look at this sequence. Draw the next two patterns in the sequence.

| 1 | 2 | 3 | 4 | 5 |

(b) Copy and complete this table to describe the sequence.

Pattern number	Red squares	White squares	Total squares
1	1	0	1
2	4	0	4
3	8	1	9
4			
5			
6			
7			

(c) Work out the formula for:

   (i) the total number of squares in pattern n

   (ii) the number of white squares in pattern n

   (iii) the number of red squares in pattern n.

2 (a) Look at this sequence. Draw the next four shapes in the sequence.

| 1 | 2 | 3 | 4 |

(b) Copy and complete this table to describe the sequence.

Pattern number	Squares up	Squares across	Total squares
1	1	2	2
2	2	3	6
3	3	4	
4	4		
5			
6			
7			
8			

(c) Work out the formula for the total number of squares in the nth pattern.

3 (a) Look at this sequence. Draw the next four patterns in the sequence.

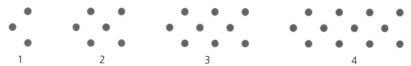

1   2   3   4

(b) Copy and complete this table to describe the sequence.

Pattern number	Number of dots
1	3
2	6
3	9
4	
5	
6	
7	
8	

(c) Work out the formula for the total number of dots in the nth pattern.

4 (a) Look at this sequence. Draw the next four patterns in the sequence.

    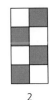

   1   2   3   4

(b) Copy and complete this table to describe the sequence.

Pattern number	Red squares	White squares	Total squares
1	2	2	4
2	4		
3	6		
4			
5			
6			
7			
8			

(c) Work out the formula for:

   (i) the total number of red squares in the nth pattern

   (ii) the total number of squares in the nth pattern.

5 (a) Look at this sequence. Draw the next four patterns in the sequence.

1 ● ●
   ● ●

2 ● ● ● ●
   ● ● ● ●

3 ● ● ● ● ● ●
   ● ● ● ● ● ●

4 ● ● ● ● ● ● ● ●
   ● ● ● ● ● ● ● ●

(b) Copy and complete this table to describe the sequence.

Pattern number	Black dots	Red dots	Total dots
1	1	3	4
2	2		
3	3		
4			
5			
6			
7			
8			

(c) Work out the formula for:

   (i) the total number of black dots in the nth pattern

   (ii) the total number of red dots in the nth pattern

   (iii) the total number of dots in the nth pattern.

6 Draw some dot or square patterns of your own.

7 Discuss your patterns with your partner. Work out formulae for the total numbers of dots or squares for each of your patterns.

Exercise 16.4: Summary exercise

1 **(a)** How many seconds are there in 1 minute?

 (b) How many seconds are there in:

   (i) 3 minutes   **(ii)** 8 minutes?

 (c) Write a formula to work out the number of seconds in m minutes.

 (d) Write a formula to work out the number of minutes in s seconds.

2 I am 3 years older than my brother. My brother is x years old. Write a formula for my age.

3 My sister is 8 years younger than I am. I am y years old.
Write a formula for my sister's age.

4 (a) At the start of the week, a plant was x cm tall. It grows 4 cm
every day. Write a formula for its height after n days.

(b) How tall will the plant be after:

    (i) 10 days

    (ii) 300 days?

Is your formula always going to be true?

5 (a) Look at this pattern and draw the next four patterns in the
sequence.

   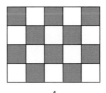

    1   2   3   4

(b) Copy and complete this table to describe the sequence.

Pattern number	Red squares	White squares	Total squares
1	1	1	2
2	3	3	6
3	6		12
4			
5			
6			
7			
8			

(c) Work out the formula for:

    (i) the total number of squares

    (ii) the number of white squares and the number of red
squares.

17 More fractions

⇨ Multiplying a fraction by a whole number

Multiplying is the same as adding over and over again.

$$3 \times \frac{3}{4} = \frac{3}{4} + \frac{3}{4} + \frac{3}{4}$$

Look at these circles, divided into quarters.

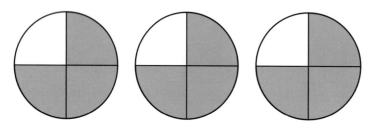

A total of 9 quarters have been shaded.

Putting the shaded quarters together to make whole circles would give:

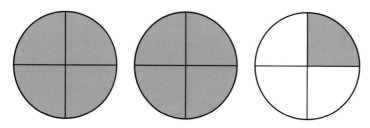

This is a total of two wholes and one quarter.

$$3 \times \frac{3}{4} = 2\frac{1}{4}$$

This can be set out as a calculation, as in these examples.

Examples:

(i) Multiply: $3 \times \frac{3}{4}$

$$3 \times \frac{3}{4} = \frac{3 \times 3}{4}$$
$$= \frac{9}{4}$$
$$= 2\frac{1}{4}$$

(ii) Multiply: $4 \times \frac{2}{5}$

$$4 \times \frac{2}{5} = \frac{4 \times 2}{5}$$
$$= \frac{8}{5}$$
$$= 1\frac{3}{5}$$

Exercise 17.1

Complete these multiplications. If the answer is an improper fraction, write it as a mixed number. Make sure that fractions in your answers are in their lowest terms.

1 $3 \times \frac{1}{4}$

2 $5 \times \frac{1}{2}$

3 $4 \times \frac{2}{7}$

4 $3 \times \frac{3}{4}$

5 $2 \times \frac{4}{5}$

6 $5 \times \frac{3}{8}$

7 $4 \times \frac{3}{7}$

8 $3 \times \frac{4}{5}$

9 $4 \times \frac{7}{8}$

10 $3 \times \frac{5}{6}$

11 $3 \times \frac{5}{12}$

12 $4 \times \frac{8}{12}$

13 $3 \times \frac{5}{16}$

14 $5 \times \frac{4}{15}$

15 $3 \times \frac{5}{7}$

16 $2 \times \frac{5}{8}$

17 $3 \times \frac{5}{9}$

18 $5 \times \frac{3}{10}$

19 $3 \times \frac{7}{15}$

20 $4 \times \frac{5}{6}$

⇨ Multiplying a mixed number by a whole number

To multiply a mixed number by a whole number, first multiply the whole number and then the fraction.

> ## Example:
> Multiply: $5 \times 1\frac{2}{3}$
>
> $$5 \times 1\frac{2}{3} = 5 \times 1 + \frac{5 \times 2}{3}$$
> $$= 5 + \frac{10}{3}$$
> $$= 5 + 3\frac{1}{3}$$
> $$= 8\frac{1}{3}$$

Exercise 17.2

Complete these multiplications. If the answer is an improper fraction, write it as a mixed number. Make sure that fractions in your answers are in their lowest terms.

1 $3 \times 1\frac{1}{4}$

2 $5 \times 2\frac{1}{7}$

3 $4 \times 1\frac{2}{9}$

4 $3 \times 3\frac{1}{5}$

5 $2 \times 2\frac{3}{8}$

6 $4 \times 2\frac{1}{8}$

7 $4 \times 3\frac{1}{4}$

8 $2 \times 3\frac{1}{4}$

9 $4 \times 1\frac{1}{12}$

10 $5 \times 2\frac{3}{20}$

11 $2 \times 1\frac{2}{3}$

12 $2 \times 2\frac{5}{7}$

13 $3 \times 2\frac{3}{4}$

14 $2 \times 1\frac{4}{5}$

15 $3 \times 2\frac{2}{5}$

16 $4 \times 2\frac{5}{6}$

17 $3 \times 1\frac{4}{9}$

18 $5 \times 3\frac{7}{10}$

19 $4 \times 2\frac{7}{12}$

20 $2 \times 3\frac{5}{8}$

⇨ Multiplying a fraction by a fraction

This rectangle is $\frac{1}{4}$ m wide and $\frac{1}{2}$ m long.

Now look at it as part of a square metre.

You can see that the rectangle is $\frac{1}{8}$ of the whole square.

$$\frac{1}{2} \times \frac{1}{4} = \frac{1}{8}$$

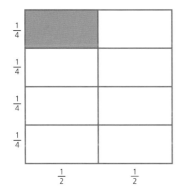

Multiplying a quarter by a half is like asking:
'What is a quarter of a half?'

Look at the fraction circles.

Here is a half.   Now divide the half into quarters.

Now you have one eighth.

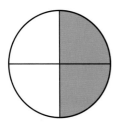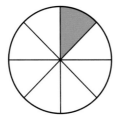

Set out the calculation like this.

$$\frac{1}{2} \times \frac{1}{4} = \frac{1}{2 \times 4}$$
$$= \frac{1}{8}$$

The method is the same, even when the fractions are not unit fractions.

Consider $\frac{3}{4} \times \frac{2}{5}$

The shaded part of this circle is $\frac{2}{5}$	This is $\frac{2}{5}$ with each $\frac{1}{5}$ having been divided into quarters.	Here are $\frac{3}{4}$ of each $\frac{1}{5}$ $\frac{3}{4}$ of $\frac{2}{5} = \frac{6}{20}$ $= \frac{3}{10}$
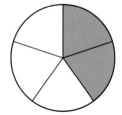	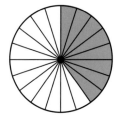	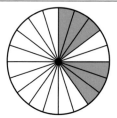

Set out the calculation like this.

$$\frac{3}{4} \times \frac{2}{5} = \frac{3}{4_2} \times \frac{2^1}{5} \qquad \text{Cancel by dividing by a common factor: 2}$$

$$= \frac{3 \times 1}{2 \times 5}$$

$$= \frac{3}{10}$$

Examples:

(i) Multiply: $\frac{1}{8} \times \frac{2}{3}$

$$\frac{1}{8} \times \frac{2}{3} = \frac{1}{8_4} \times \frac{2^1}{3} \qquad \text{Cancel by dividing by a common factor: 2}$$

$$= \frac{1 \times 1}{4 \times 3}$$

$$= \frac{1}{12}$$

(ii) Multiply: $\frac{3}{8} \times \frac{4}{9}$

$$\frac{3}{8} \times \frac{4}{9} = \frac{3^1}{8_2} \times \frac{4^1}{9_3} \qquad \text{Cancel by dividing by common factors: 4 and 3}$$

$$= \frac{1 \times 1}{2 \times 3}$$

$$= \frac{1}{6}$$

Exercise 17.3

Complete these multiplications. If the answer is an improper fraction, write it as a mixed number. Make sure that fractions in your answers are in their lowest terms.

1 $\frac{1}{3} \times \frac{1}{4}$

2 $\frac{1}{4} \times \frac{1}{5}$

3 $\frac{1}{5} \times \frac{1}{2}$

4 $\frac{1}{3} \times \frac{1}{5}$

5 $\frac{1}{2} \times \frac{1}{8}$

6 $\frac{1}{4} \times \frac{1}{8}$

7 $\frac{1}{6} \times \frac{1}{4}$

8 $\frac{1}{5} \times \frac{1}{3}$

9 $\frac{1}{2} \times \frac{1}{7}$

10 $\frac{1}{6} \times \frac{1}{5}$

11 $\frac{1}{2} \times \frac{2}{3}$

12 $\frac{1}{4} \times \frac{2}{7}$

13 $\frac{1}{3} \times \frac{3}{4}$

14 $\frac{1}{2} \times \frac{4}{5}$

15 $\frac{1}{4} \times \frac{8}{9}$

16 $\frac{1}{5} \times \frac{5}{6}$

17 $\frac{1}{8} \times \frac{6}{7}$

18 $\frac{1}{3} \times \frac{9}{10}$

19 $\frac{1}{2} \times \frac{6}{7}$

20 $\frac{1}{6} \times \frac{9}{11}$

21 $\frac{5}{6} \times \frac{2}{3}$

22 $\frac{3}{4} \times \frac{2}{9}$

23 $\frac{2}{3} \times \frac{3}{4}$

24 $\frac{5}{6} \times \frac{4}{5}$

25 $\frac{3}{4} \times \frac{5}{6}$

26 $\frac{3}{8} \times \frac{2}{9}$

27 $\frac{7}{8} \times \frac{4}{7}$

28 $\frac{2}{3} \times \frac{9}{10}$

29 $\frac{3}{5} \times \frac{5}{12}$

30 $\frac{8}{9} \times \frac{3}{4}$

⇨ Multiplying a mixed number by a fraction

Consider $3\frac{1}{2} \times \frac{2}{3}$

You could find $3 \times \frac{2}{3} = 2$ and then $\frac{1}{2} \times \frac{2}{3} = \frac{1}{3}$ and add them together to get $2\frac{1}{3}$ but it is much easier to do the calculation in one step, by turning the mixed number into an **improper fraction**.

Examples:

(i) Multiply: $3\frac{1}{2} \times \frac{2}{3}$

$$3\frac{1}{2} \times \frac{2}{3} = \frac{7}{2_1} \times \frac{2^1}{3}$$

$$= \frac{7 \times 1}{1 \times 3}$$

$$= \frac{7}{3}$$

$$= 2\frac{1}{3}$$

(ii) Multiply: $\frac{5}{8} \times 1\frac{1}{7}$

$$\frac{5}{8} \times 1\frac{1}{7} = \frac{5}{8_1} \times \frac{8^1}{7}$$

$$= \frac{5 \times 1}{1 \times 7}$$

$$= \frac{5}{7}$$

(iii) Multiply: $\frac{4}{9} \times 1\frac{7}{8}$

$$\frac{4}{9} \times 1\frac{7}{8} = \frac{4^1}{9_3} \times \frac{15^5}{8_2}$$

$$= \frac{1 \times 5}{3 \times 2}$$

$$= \frac{5}{6}$$

Exercise 17.4

Complete these multiplications. If the answer is an improper fraction, write it as a mixed number. Make sure that fractions in your answers are in their lowest terms.

1 $1\frac{1}{3} \times \frac{1}{4}$

2 $1\frac{1}{4} \times \frac{1}{5}$

3 $\frac{1}{5} \times 2\frac{1}{2}$

4 $3\frac{1}{3} \times \frac{1}{5}$

5 $4\frac{1}{2} \times \frac{1}{9}$

6 $1\frac{1}{5} \times \frac{1}{3}$

7 $4\frac{1}{2} \times \frac{2}{3}$

8 $\frac{1}{2} \times 2\frac{2}{3}$

9 $\frac{1}{8} \times 3\frac{3}{5}$

10 $1\frac{1}{6} \times \frac{3}{7}$

11 $\frac{4}{9} \times 2\frac{5}{8}$

12 $\frac{3}{4} \times 2\frac{2}{7}$

13 $5\frac{1}{3} \times \frac{3}{4}$

14 $\frac{4}{5} \times 2\frac{1}{7}$

15 $2\frac{11}{12} \times \frac{3}{7}$

16 $\frac{4}{5} \times 5\frac{5}{6}$

17 $\frac{3}{5} \times 4\frac{1}{6}$

18 $3\frac{3}{4} \times \frac{2}{5}$

19 $\frac{4}{7} \times 3\frac{1}{2}$

20 $5\frac{5}{6} \times \frac{3}{7}$

⇨ Dividing a whole number by a fraction

How many halves are there in three wholes?

Each of these circles has been divided into halves. There are six halves altogether.

$$3 \div \frac{1}{2} = 6$$

When you divide a whole number by a fraction that is less than one, the answer is **larger** than the original number.

Think about dividing as sharing. If you had five sandwiches and wanted to give each child half a sandwich, how many children would get half a sandwich?

$$5 \div \frac{1}{2} = 10$$

That is the same as saying $5 \div \frac{1}{2} = 5 \times 2 = 10$

So you could give 10 children half a sandwich.

Look again at what is happening.

You know that there are 2 halves in 1 whole.

So in 5 wholes, there are 5 × 2 halves.

The fraction has turned upside down and the ÷ has become ×.

Why does that happen?

Examples:

(i) Divide: $5 \div \frac{1}{3}$

$$5 \div \frac{1}{3} = 5 \times \frac{3}{1}$$

$$= \frac{5 \times 3}{1}$$

$$= 15$$

(ii) Divide: $5 \div \frac{3}{4}$

$$5 \div \frac{3}{4} = 5 \times \frac{4}{3}$$

$$= \frac{5 \times 4}{3}$$

$$= \frac{20}{3}$$

$$= 6\frac{2}{3}$$

(iii) Divide: $4 \div \frac{2}{3}$

$$4 \div \frac{2}{3} = 4^2 \times \frac{3}{2_1}$$

$$= \frac{2 \times 3}{1}$$

$$= 6$$

Have you worked it out?

$\frac{1}{3}$ is $1 \div 3$

The inverse of ÷ is ×

Therefore $\div \frac{1}{3}$ is the same as $\div 1 \times 3 = \frac{3}{1}$

And $\div \frac{3}{4}$ is the same as $\div 3 \times 4 = \frac{4}{3}$

Exercise 17.5

Complete these divisions. If the answer is an improper fraction, write it as a mixed number. Make sure that fractions in your answers are in their lowest terms.

1 $1 \div \frac{1}{4}$

2 $2 \div \frac{1}{5}$

3 $3 \div \frac{1}{3}$

4 $3 \div \frac{1}{5}$

5 $4 \div \frac{1}{9}$

6 $5 \div \frac{1}{3}$

7 $6 \div \frac{1}{7}$

8 $4 \div \frac{1}{3}$

9 $3 \div \frac{1}{6}$

10 $5 \div \frac{1}{4}$

11 $5 \div \frac{5}{8}$

12 $4 \div \frac{2}{5}$

13 $6 \div \frac{3}{4}$

14 $6 \div \frac{3}{7}$

15 $4 \div \frac{6}{7}$

16 $5 \div \frac{10}{11}$

17 $15 \div \frac{6}{7}$

18 $12 \div \frac{3}{4}$

19 $16 \div \frac{8}{9}$

20 $15 \div \frac{9}{10}$

⇨ Dividing a fraction by a fraction

You have seen that to divide a whole number by a fraction, you turn the fraction upside down and change the division to a multiplication.

The method for dividing a fraction or a mixed number by a fraction is the same. Turn the fraction you are dividing by upside down, then multiply.

$$\frac{1}{3} \div \frac{1}{2} = \frac{1}{3} \times \frac{2}{1}$$
$$= \frac{1 \times 2}{3 \times 1}$$
$$= \frac{2}{3}$$

Examples:

(i) Divide: $\frac{1}{5} \div \frac{1}{2}$

$$\frac{1}{5} \div \frac{1}{2} = \frac{1}{5} \times \frac{2}{1}$$

$$= \frac{1 \times 2}{5 \times 1}$$

$$= \frac{2}{5}$$

(ii) Divide: $\frac{3}{5} \div \frac{3}{4}$

$$\frac{3}{5} \div \frac{3}{4} = \frac{3^1}{5} \times \frac{4}{3_1}$$

$$= \frac{1 \times 4}{5 \times 1}$$

$$= \frac{4}{5}$$

(iii) Divide: $4\frac{2}{3} \div \frac{7}{9}$

$$4\frac{2}{3} \div \frac{7}{9} = \frac{14}{3} \div \frac{7}{9}$$

$$= \frac{14^2}{3_1} \times \frac{9^3}{7_1}$$

$$= \frac{2 \times 3}{1 \times 1}$$

$$= 6$$

Exercise 17.6

Complete these divisions. If the answer is an improper fraction, write it as a mixed number. Make sure that fractions in your answers are in their lowest terms.

1 $\frac{1}{3} \div \frac{1}{4}$

2 $\frac{1}{7} \div \frac{1}{5}$

3 $\frac{1}{4} \div \frac{1}{3}$

4 $\frac{1}{3} \div \frac{1}{6}$

5 $\frac{1}{6} \div \frac{1}{9}$

6 $\frac{1}{4} \div \frac{1}{6}$

7 $\frac{1}{4} \div \frac{1}{12}$

8 $\frac{1}{12} \div \frac{1}{3}$

9 $\frac{1}{12} \div \frac{1}{9}$

10 $\frac{1}{8} \div \frac{1}{12}$

11 $\frac{3}{4} \div \frac{5}{8}$

12 $\frac{3}{5} \div \frac{2}{5}$

13 $\frac{3}{5} \div \frac{3}{4}$

14 $\frac{2}{5} \div \frac{4}{7}$

15 $\frac{3}{10} \div \frac{6}{7}$

16 $\frac{5}{6} \div \frac{2}{3}$

17 $\frac{9}{14} \div \frac{6}{7}$

18 $\frac{9}{20} \div \frac{3}{4}$

19 $\frac{5}{16} \div \frac{5}{8}$

20 $\frac{3}{5} \div \frac{9}{10}$

21 $1\frac{3}{4} \div \frac{7}{8}$

22 $1\frac{3}{5} \div \frac{2}{5}$

23 $3\frac{3}{5} \div \frac{3}{10}$

24 $2\frac{2}{5} \div \frac{3}{10}$

25 $3\frac{3}{10} \div \frac{11}{12}$

26 $2\frac{11}{12} \div \frac{5}{6}$

27 $2\frac{1}{4} \div \frac{3}{8}$

28 $1\frac{4}{5} \div \frac{3}{10}$

29 $1\frac{5}{9} \div \frac{7}{12}$

30 $2\frac{3}{15} \div \frac{9}{10}$

Exercise 17.7: Summary exercise

Complete these multiplications and divisions. If the answer is an improper fraction, write it as a mixed number. Make sure that fractions in your answers are in their lowest terms.

1 $3 \times \frac{1}{4}$

2 $4 \times \frac{1}{2}$

3 $8 \times \frac{3}{4}$

4 $4 \times 1\frac{1}{8}$

5 $3 \times 2\frac{1}{6}$

6 $\frac{3}{5} \times \frac{1}{2}$

7 $\frac{2}{3} \times \frac{7}{8}$

8 $\frac{1}{3} \times 1\frac{1}{4}$

9 $\frac{9}{10} \times 4\frac{1}{6}$

10 $5 \div \frac{1}{8}$

11 $9 \div \frac{6}{7}$

12 $\frac{1}{4} \div \frac{1}{5}$

13 $2\frac{1}{4} \div \frac{15}{16}$

14 $1\frac{1}{5} \div \frac{3}{4}$

15 In my class, $\frac{1}{2}$ the class are girls and $\frac{1}{3}$ of the girls have at least one brother. What fraction of the class are girls with at least one brother?

16 Sam ate $\frac{1}{2}$ a pizza on Friday. On Saturday he ate $\frac{1}{3}$ of what was left and gave the rest to the dog. What fraction of the pizza did the dog have?

17 The zoo ordered $\frac{9}{10}$ of a barrel of monkey nuts. They gave $\frac{2}{3}$ of the nuts to the chimpanzee and the rest to the orangutan. What fraction of a barrel of nuts did the chimpanzee and the orangutan each have?

18 I have $\frac{5}{8}$ of a litre of cordial to pour equally into 6 glasses. What fraction of a litre goes in each glass?

19 A recipe for 36 biscuits needs $1\frac{2}{3}$ cups of sugar. I am going to make 24 biscuits, how many cups of sugar do I need?

20 A farmer orders $3\frac{3}{4}$ bales of hay. He spreads $\frac{2}{5}$ of the delivery around the stables and stores the rest. How many bales of hay are stored?

18 Time and travel

⇨ Time

Time is an interesting measurement. Although it is so much part of our lives, it can be hard to measure or estimate.

Try closing your eyes and then opening them after one minute. How accurate were you?

⇨ Units of time

$$60 \textbf{ seconds} = 1 \textbf{ minute}$$

$$60 \textbf{ minutes} = 1 \textbf{ hour}$$

$$24 \textbf{ hours} = 1 \textbf{ day}$$

$$7 \textbf{ days} = 1 \textbf{ week}$$

$$28–31 \textbf{ days} = 1 \textbf{ month}$$

$$365 \textbf{ days} = 1 \textbf{ year}$$

$$366 \textbf{ days} = 1 \textbf{ leap year}$$

$$12 \textbf{ months} = 1 \textbf{ year}$$

The units of time for seconds, minutes, hours and days are always the same. There are always 60 minutes in an hour, there are always 24 hours in one day.

This is not true for months and years. Although a month is based on the time it takes the Moon to go round the Earth (a **lunar month**), which is about $29\frac{1}{2}$ days, the actual lengths of the months vary, for historical reasons.

This rhyme will help you to remember the rules.

> *Thirty days has September,*
> *April, June and November.*
> *All the rest have thirty-one*
> *Except for February alone*
> *And that has twenty-eight days clear*
> *and twenty-nine in any leap year.*

A year is the time it takes for the Earth to travel round the Sun. This is almost exactly $365\frac{1}{4}$ days. To take account of these extra quarters of days, there is an extra day in every fourth year (a **leap year**).

Leap years occur when the year number is a multiple of 4: 2004, 2016, 1816. However, to make the time work out exactly, the turn of the century years are not leap years unless they are multiples of 400. 2000 was leap year but 1900 was not.

Exercise 18.1

1 How many minutes are there in a day?

2 How many hours are there in a week?

3 How many weeks and days are there in a 'common' year (i.e. not a leap year)?

4 In 2014 my birthday falls on a Tuesday. On what day does it fall in 2015?

5 How many weeks and days are there in a leap year?

6 If 1 May is a Friday in 2015, what day of the week will it be in 2016?

7 How many days are there from 1 February 2016 to 31 May 2016?

8 It is going to take the builders 40 weekdays to mend our roof. If they start on Monday 5 May, when will they finish?

9 The school term starts on 14 September 2016 and finishes on 18 December. How many days is that?

10 Queen Elizabeth II ascended to the throne on 6 February 1952 How many days ago is that?

⇨ The 24-hour clock

We talk about the time in three ways.

Using past and to:

> Half past four
>
> Ten to three

Using a.m. (morning) and p.m. (afternoon and evening)

> I get up at 7.15 a.m.
>
> I go to bed at 8.30 p.m.

Using 24-hour clock times

> The bus arrives at 07:20
>
> The train leaves at 16:45

> ✳ It is usual to separate the minutes from the hours, in a 24-hour time, by a colon, although in some timetables they are closed up.

Exercise 18.2

1 Write the times on these clocks as 24-hour clock times.

(a) Morning times

(i)   (ii)   (iii)

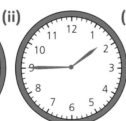

(b) Afternoon and evening times

(i)   (ii)   (iii)

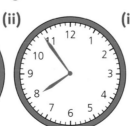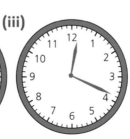

2 You will need a clock like this to answer each part of this question.

Draw the hands on your clocks to show these times:

(a) A quarter past three

(b) 7.25 a.m.

(c) 1955

(d) Twenty-five to six

(e) 4.28 p.m.

(f) 13:14

(g) Seventeen minutes past nine

(h) 00.06 a.m.

(i) 12:44

(j) Ten to ten

3 Write these 24-hour clock times as 'past' or 'to' times.

(a) 14:15

(b) 07:45

(c) 13:30

(d) 06:17

(e) 23:55

(f) 00:12

(g) 14:06

(h) 12:34

(i) 06:41

(j) 16:29

4 A train leaves London at 14:44 and arrives in Birmingham at 16:09
How long does the journey take?

5 A train leaves Manchester at 02:19 and arrives in Glasgow at 10:37
How long does the journey take?

6 A train leaves Southampton at 19:19 and arrives in Exeter at 22:12
How long does the journey take?

7 I need to travel from Bristol to Inverness. I have to catch three trains. These are the times I have been given.

Leave Bristol   07:39   Arrive Edinburgh   14:10

Leave Edinburgh   14:35   Arrive Perth   15:53

Leave Perth   16:16   Arrive Inverness   18:21

(a) How long do I have to wait at Edinburgh?

(b) How long do I have to wait at Perth?

(c) How long does it take for my total journey?

(d) How much time will I spend on the trains?

8 We are planning a holiday by ferry to France. These are the possible routes and the times they take.

	Leaves UK	Time taken
Dover to Calais	07:55	1 hours 12 minutes
Portsmouth to Caen	08:25	6 hours 33 minutes
Poole to Cherbourg	08:45	4 hours 19 minutes

Work out the arrival time of each ferry.

9 These ferries go to Spain. That is a long way and so the ferries arrive the day after they set out.

	Depart	Arrive
Portsmouth to Bilbao	15:45	11:15
Plymouth to Santander	17:35	16:15

(a) Work out the time taken for each journey.

(b) The ferry below takes two nights for the journey. Work out the time it takes to make the passage.

	Depart	Arrive
Portsmouth to Santander	22:30	06:45

⇨ Time as a fraction

When calculating with time it can be helpful to write hours and minutes as fractions of hours.

There are 60 minutes in an hour and 60 has lots of factors. This means that you can usually simplify the fraction and then calculate with simple numbers.

Example:

Write these times as fractions of hours.

(i) 2 hours 30 minutes   (ii) 4 hours 40 minutes   (iii) 6 hours 35 minutes

(i) 2 hours 30 minutes $= 2\frac{30}{60} = 2\frac{1}{2}$ hours

(ii) 4 hours 40 minutes $= 4\frac{40}{60} = 4\frac{2}{3}$ hours

(iii) 6 hours 35 minutes $= 6\frac{35}{60} = 6\frac{7}{12}$ hours

Exercise 18.3

1 Write these times as fractions of hours.

(a) 30 minutes

(b) 15 minutes

(c) 1 hour 45 minutes

(d) 3 hours 10 minutes

(e) 4 hours 20 minutes

(f) 5 hours 12 minutes

(g) 4 hours 50 minutes

(h) 1 hour 5 minutes

(i) 2 hours 25 minutes

(j) 5 hours 36 minutes

2 Write these times as hours and minutes.

(a) $1\frac{1}{2}$ hours

(f) $3\frac{3}{4}$ hours

(b) $3\frac{1}{4}$ hours

(g) $4\frac{1}{6}$ hours

(c) $\frac{1}{3}$ hour

(h) $2\frac{5}{6}$ hours

(d) $\frac{1}{12}$ hour

(i) $1\frac{3}{5}$ hours

(e) $2\frac{2}{3}$ hours

(j) $2\frac{5}{12}$ hours

⇨ Speed, distance and time

In Exercise 18.2 you worked out the time taken for a journey by considering the departure time and the arrival time. The time taken for any journey will depend on two things, the distance you are travelling and the speed at which you travel.

The **faster** you are going the **shorter** your journey time will be.

Exercise 18.4

1 These are the finishing times, in minutes and seconds, for the five competitors in a race. List them in finishing order, starting with the winner.

Name	Time
Ali	12 minutes 20 seconds
Bill	12 minutes 14 seconds
Charles	11 minutes 49 seconds
Dee	11 minutes 55 seconds
Ed	12 minutes 5 seconds

2 Three cars are travelling on the motorway from London to Bristol. The blue car is travelling at 60 miles per hour, the red car at 70 miles per hour and the yellow car at 67 miles per hour. Which car has the shortest journey time?

3 My school is twice as far away as my sister's school.

   (a) If we walk to school at the same speed, who arrives there first?

   (b) I decide to run. I can run twice as fast as my sister can walk. Who arrives at school first when I run?

   (c) I am going to cycle to school. I can cycle five times faster than my sister can walk. Who arrives at school first now?

4 We did a class survey of the time it took us to get to school. These are some of the results.

	Distance	Time
Anna	200 metres	5 minutes
Ben	1.5 km	6 minutes
Carla	1.5 km	25 minutes
Denzil	5 km	10 minutes

Work out who came to school by car, who cycled and who walked.

⇨ Calculating speed

How good are you at judging how fast things travel? It helps to have some idea, so that you can check that your answers are sensible.

You will have seen signs on roads like this.

This means that there is a speed limit of 30 miles per hour.

In the UK, distance is measured in miles and therefore speed is calculated in **miles per hour** (mph).

In Europe, distance is measured in kilometres and therefore speed is calculated in **kilometres per hour** (kph).

As a close approximation, 8 km ≈ 5 miles.

Then 80 kph ≈ 50 mph.

The simplest way to calculate speed is to work out the distance travelled in one hour.

If you travel 60 km in one hour, your speed is 60 km per hour (or 60 kph).

If you travel 60 km in two hours, then you would travel 30 km in one hour and your speed is 30 km per hour (30 kph).

If you travel 60 km in 30 minutes, you would travel 120 km in one hour and your speed is 120 km per hour (120 kph).

Examples:

(i) A bus travels 15 miles in 20 minutes. What is its speed?

It travels 15 miles in 20 minutes

It travels 15 × 3 miles in 20 × 3 minutes   Multiply both 15 and 20 by 3

It travels 45 miles in 60 minutes

Its speed is 45 mph

(ii) A car travelled 320 km in 4 hours. What was its speed?

It travelled 320 km in 4 hours

It travelled 80 km in 1 hour   Divide both 320 and 4 by 4

Its speed was 80 kph

(iii) I walk 1 km in 12 minutes. What is my speed?

I walk 1 km in 12 minutes

I walk 5 km in 60 minutes   Multiply both 1 and 12 by 5

My speed is 5 kph

1 Measure a distance of 100 metres.

   Time yourself walking 100 metres and then multiply this time by 10
   That is the time it takes you to walk 1 km. Calculate your speed in
   kilometres per hour.

2 A car travels 70 km in one hour. What is its speed?

3 Faisal runs 5 km in half an hour. What is his speed?

4 A cyclist travels 10 miles in 20 minutes. What is her speed?

5 A car travels 15 miles in 15 minutes. What is its speed?

6 A boat completes a passage of 65 miles in 5 hours. What is its speed?

7 A plane travels 3200 km in four hours. What is its speed?

8 A train travels 200 km in two and a half hours. What is its speed?

9 I walk 500 m in 10 minutes. What is my speed?

10 A sprinter ran 100 m in 10 seconds. What was his speed?

11 A car travels 2 km in 5 minutes. What is its speed?

12 **(a)** Copy this table. Add another column and head it 'Speed'.
   Calculate the speed of each person.

	Distance	Time
Anna	200 metres	5 minutes
Ben	1.5 km	6 minutes
Carla	1.5 km	25 minutes
Denzil	5 km	10 minutes

   (b) Put Anna, Ben, Carla and Denzil in order, fastest first.

⇨ Calculating distance

When you know the speed, you know how far something can travel
in one hour. From that you can multiply, to work out how far it can
go in more than one hour. You will need to turn times that are less
than an hour into **fractions** and then multiply by the fraction.

Examples:

(i) A car is travelling at 40 miles per hour. How far does it travel in an hour and a half?

Distance $= 40 \times 1\frac{1}{2}$

$= 60$ miles

(ii) Eva cycles at 25 kph for 15 minutes. How far does she travel?

Distance $= 25 \times \frac{1}{4}$

$= 6\frac{1}{4}$ km

Exercise 18.6

1 George walks at 4 mph for 2 hours. How far does he go?

2 My brother cycles to school in 15 minutes. If his speed is 32 mph, how far is it to his school?

3 Calculate the distances of these journeys.

	Speed	Time
(a)	10 kph	$1\frac{1}{2}$ hours
(b)	55 mph	2 hours
(c)	800 kph	$3\frac{1}{2}$ hours
(d)	30 mph	45 minutes
(e)	45 mph	12 minutes

Which of (a), (b), (c), (d) and (e) would be an aeroplane?

4 A car is travelling at 65 mph. It takes 3 hours to go from London to Leeds. What is the distance travelled?

5 It took us an hour and a half to get to Dover. We were travelling at 60 miles per hour. How far did we have to go?

6 It takes the ferry 1 hour 20 minutes to sail from Dover to Calais at 36 mph. How far is it from Dover to Calais?

7 It takes us 3 hours 20 minutes to drive from Calais to Paris at 90 km per hour. How far is it from Calais to Paris?

⇨ Calculating time

If you know the distance and the speed, then you can use the same method of multiplying or dividing to find the time taken.

Examples:

(i) A car is travelling at 60 miles per hour. How long does it take to travel 90 miles?

It travels 60 miles in 1 hour

It travels 30 miles in $\frac{1}{2}$ hour

It travels 90 miles in $1\frac{1}{2}$ hours

You can see that 30 is the highest common factor of 60 and 90

(ii) I can cycle at 40 km per hour. How long will it take me to cycle 90 km?

I can cycle 40 km in 1 hour

I can cycle 10 km in $\frac{1}{4}$ hour

I can cycle 90 km in $9 \times \frac{1}{4} = 2\frac{1}{4}$ hours

Exercise 18.7

1 How long does it take to walk 8 km at a speed of 4 kph?

2 How long does it take a car to drive 75 miles at a speed of 60 mph?

3 Calculate the times taken to complete these journeys.

	Speed	Distance
(a)	20 kph	5 km
(b)	40 mph	100 miles
(c)	100 kph	330 km

Describe some possible journeys that could be made by (a), (b) and (c).

4 It is 12 miles from Bath to Bristol. How long does it take a bus travelling at 30 miles an hour to travel from Bristol to Bath?

5 A plane leaves Heathrow at 08:30
It travels at 400 mph. What time will it arrive in Madrid,
900 miles away?

6 A train leaves Paddington at 09:55 and travels to Crewe, at
60 miles an hour. What time will it arrive at Crewe, 170 miles
away?

7 (a) How long does it take me to cycle 5 km to school, at a speed of
30 km per hour?

(b) If I have to be at school by 08:30, what time must I leave home?

8 Copy and complete this table of travel times.

	Depart	Speed	Distance	Arrive
(a)	07:10	40 kph	16 km	
(b)	08:20	60 mph	90 miles	
(c)		500 mph	1200 miles	01:10

⇨ Using a formula

In Chapter 16, you worked out the formulae for distance, speed and
time:

Distance = speed × time

$$d = s \times t$$

Time = distance ÷ speed

$$t = \frac{d}{s}$$

Speed = distance ÷ time

$$s = \frac{d}{t}$$

When using these formulae, make sure that you have the correct
units for time. If you are working in miles per hour, your time must
be in hours not minutes.

Examples:

(i) How long does it take a lorry travelling at 60 mph to travel 84 miles?

$$\text{Time} = \frac{d}{s}$$

$$= \frac{84}{60} \text{ hours}$$

$$= 1 \text{ hour 24 minutes}$$

(ii) How far does a car travelling at 50 mph go in 1 hour 12 minutes?

First, change the time to a fraction.

$$\text{Distance} = s \times t$$

$$= 5\cancel{0} \times \frac{\cancel{72}^{12}}{\cancel{60}_{1}}$$

$$= 60 \text{ miles}$$

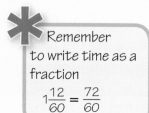

Remember
to write time as a
fraction

$$1\frac{12}{60} = \frac{72}{60}$$

(iii) I cycle 12 km in 45 minutes. What is my speed?

First, change the time to hours.

45 minutes is $\frac{45}{60}$ hours.

$$\frac{45}{60} = \frac{3}{4}$$

$$\text{Speed} = \frac{d}{t}$$

$$= 12 \div \frac{3}{4}$$

$$= \frac{\cancel{12}^{4} \times 4}{\cancel{3}_{1}} \qquad \text{Divide 12 and 3 by 3}$$

$$= 16 \text{ km per hour (kph)}$$

Exercise 18.8

Use the correct formula to solve these problems. Remember to show all your working and set your work out carefully.

1 How long does it take a rambler to travel a distance of 14 km at 5 kph?

2 How far do I go if I travel at 36 mph for an hour and a quarter?

3 A train travels 130 miles in 2 hours 10 minutes. What is its speed?

4 Calculate the missing values in this table.

Distance	Speed	Time
(a) ...	88 kph	$\frac{1}{2}$ hour
180 miles	(b) ...	2 hours 15 minutes
28 miles	21 mph	(c) ...

5 In 2013, Paula Radcliffe set a women's record for running a marathon of 26 miles in 2 hours 15 minutes. What was her speed?

6 I catch a train at half past nine. The train travels for 120 km at a speed of 50 km per hour.

(a) How long does the journey take?

(b) At what time do I arrive?

7 I land in New York at 14:20 in a plane that has flown at 500 mph for 3450 miles. At what time did the plane leave?

8 This is my flight itinerary.

Depart London 08:45   Time to Vienna: 1 hour 42 minutes

Distance from London to Vienna: 765 miles

Depart Vienna 12:05   Time to Cairo: 3 hours 05 minutes

Distance from Vienna to Cairo: 1480 miles

(a) Which was the faster plane?

(b) Time in Vienna is one hour ahead of London time.

   (i)   What time did I arrive in Vienna, local time?

   (ii)   How long did I have to wait for my next flight?

(c) Time in Cairo is one hour ahead of Vienna time. What time did I arrive in Cairo, local time?

19 Line graphs

1/5/2020

⇨ What is a line graph?

There are two main types of line graph.

- A line graph can show how variable data changes over time.

- A line graph can also show a direct relationship, such as a currency conversion.

⇨ Line graphs to show how a variable changes over time

You will find examples of graphs that show how a **variable** and **measurable** data changes, over time, in your History, Geography and Science textbooks.

You can read values directly from such a graph. You can also deduce information that helps you understand more about what you are studying.

This graph from a History textbook shows the population figures for Britain from 1100 to 2000. You can see how the population has grown from under 2 million in 1100 to nearly 50 million in the year 2000

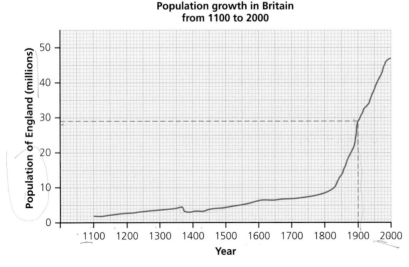

Population growth in Britain from 1100 to 2000

Axis x – time

Now – 67 million 2020

To read a value from the graph you can draw a horizontal line and a vertical line from the graph to the axes. The lines drawn on this graph show that in 1900 the population in Britain was about 29 million.

Scale

Before you can read information from a graph, you must make sure that you understand the scale.

On this graph, ten little squares represent 1 hour, one little square will be 6 minutes.

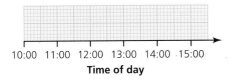

Time of day

On this graph, ten little squares represent 20 cm, one little square will be 2 cm.

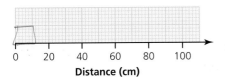

Distance (cm)

Exercise 19.1

1 Look at the scale on this graph.

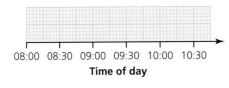

Time of day

(a) How many minutes does one small square represent? 3

(b) How many small squares will represent a time interval of:

    (i) 24 minutes 8   (ii) 3 hours? $\dfrac{180}{3}$ 60 = 1 hr
                                                     180 = 3 hr

= 60 small square.

2 Look at the scale on this graph.

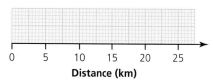

0   5   10   15   20   25

Distance (km)

(a) What distance does one small square represent? 0.5

(b) How many small squares will represent an interval of:

   (i) 1 km 2   **(ii)** 12 km? 24

3 Look at the scale on this graph.

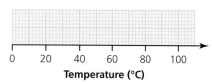

0   20   40   60   80   100

Temperature (°C)

(a) How many degrees does one small square represent? 2⁰

(b) How many small squares will represent a temperature difference of:

   (i) 10 degrees 5   **(ii)** 5 degrees? 2.5

4 Look at the scale on this graph.

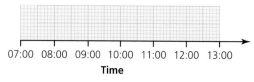

07:00 08:00 09:00 10:00 11:00 12:00 13:00

Time

(a) How many minutes does one small square represent? 6

(b) How many small squares will represent a time difference of:

   (i) 10 minutes 1.6   **(ii)** 20 minutes?

If 7.10 → 7.20

   10 minutes.

   $\frac{6}{6}$ 1$\frac{4}{6}$ squares.

20 minuts

   $\frac{20}{6}$ = 3$\frac{2}{6}$.

   = 3 1/3 squars

5 Look again at the graph of the population of Britain.

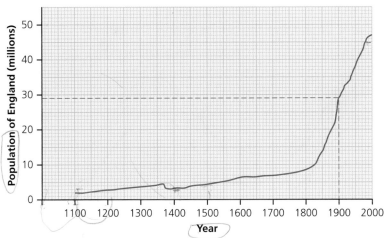

**Population growth in Britain
from 1100 to 2000**

Population of England (millions) — vertical axis
Year — horizontal axis

(a) (i) What period of time does one small square on the horizontal axis represent? *10 yrs*

(ii) What population does one small square on the vertical axis represent? *1m people*

(b) Look at the date where the graph starts. What year do you think it is and why? *1000*

(c) There is a dip in population between 1300 and 1400 Why do you think that was? *The Plague*

(d) What was the approximate rise in population from:

(i) 1400 to 1500 *1m people*

(ii) 1500 to 1600 *2m people*

(iii) 1600 to 1700 *1m people*

(iv) 1700 to 1800 *2m people*

(v) 1800 to 1900 *20m people*

(vi) 1900 to 2000? *18m people*

(e) What might explain the slowness of the growth between 1600 and 1700?

(f) When does the rapid rise in population start? *1800*

Give as many reasons for this as you can.

6 These two graphs show the temperatures in Manchester and in Reykjavik.

The red line is the daytime temperature and the blue line is the night-time temperature.

(a) The two graphs look very similar. Are the temperatures the same? *No*

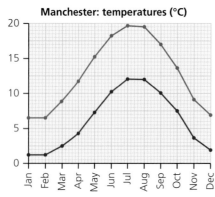
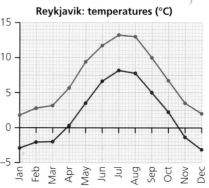

(b) What is the highest daytime temperature in:

(i) Manchester *19.5 c°* **(ii)** Reykjavik? *12.5 c°*

(c) What is the lowest night-time temperature in:

(i) Manchester *1 c°* **(ii)** Reykjavik? *-3 c°*

(d) When is the biggest difference between night-time and daytime temperatures and in which city?

7 Look at this temperature graph for Heraklion, in Crete, Greece.

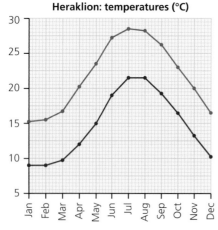

(a) Look at a map of Europe and find the three cities. Describe the locations of Heraklion and Reykjavik.

(b) Compare the differences in the temperatures in the two places.

> ✳ Do not just refer to them as 'hotter' or 'cooler', compare the actual temperatures.

8 This climate graph includes a bar chart for rainfall against a line graph for average temperature.

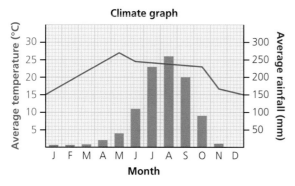

Climate graph

(a) What is the highest temperature recorded?

(b) What month has the most rainfall?

(c) In which months is the average temperature 22 °C?

(d) In which month was the rainfall 200 mm?

(e) Describe the climate between the months of November and March.

(f) Describe the climate between the months of May and October.

(g) Do you think the location for this climate graph is desert, temperate, savannah or tropical rainforest?

9 This graph shows a car journey on a shopping trip. It has time on the horizontal axis and distance on the vertical axis.

The graph shows that the car is driven quite quickly after leaving the house but then more slowly for a while, before stopping. After some time at the shops, the car is then driven home at a steady pace.

A shopping trip

(a) How many minutes does one small square on the horizontal axis represent?

(b) How many small squares on the vertical axis represent 0.5 km?

(c) For how long does the car travel before it slows down?

(d) How long is the car stationary while the driver goes shopping?

(e) How far from home is the car at:

(i) 10:30 **(ii)** 11:50?

(f) At what two times is the car 7 km from home?

(g) What time is it when the car arrives home?

(h) Between what times is the car going fastest? How can you tell, from the graph?

Drawing line graphs
Here are some things to remember when you are discussing or drawing graphs.

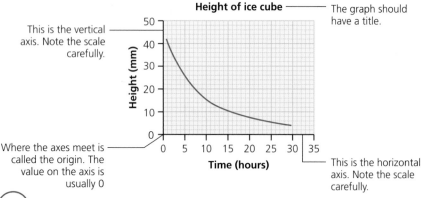

Height of ice cube — The graph should have a title.

This is the vertical axis. Note the scale carefully.

Where the axes meet is called the origin. The value on the axis is usually 0

This is the horizontal axis. Note the scale carefully.

Exercise 19.2

You will need graph paper for this exercise.

1   In Science we measured the temperature of a hot drink as it cooled over an hour. Here are our results.

Time (minutes)	Temperature (°C)	Time (minutes)	Temperature (°C)
0	90	35	22
5	75	40	20
10	62	45	20
15	48	50	20
20	40	55	20
25	31	60	20
30	25		

(a) Draw a set of axes on graph paper. Label the horizontal axis 'Time (minutes)' and mark it in 5-minute intervals. Label the vertical axis 'Temperature (°C)' and mark it in 10-degree intervals.

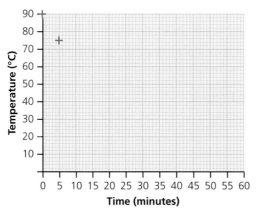

(b) Carefully mark a small cross (+) at each five-minute interval at the correct temperature. The first two are shown on the graph above.

(c) Now join the crosses with a smooth curve.

(d) During which five-minute period was the temperature drop the greatest? How is that shown on the graph?

(e) When was the temperature drop the smallest? How is that shown on the graph?

(f) What do you think the room temperature might have been? Why?

2 This table shows the growth of three plants over 12 days.

Time (days)	Growth (cm)		
	Plant A	Plant B	Plant C
0	1.0	1.5	1.0
2	1.5	2.0	2.2
4	2.2	2.7	3.7
6	3.2	3.2	5.4
8	4.3	3.5	7.0
10	5.2	3.7	8.7
12	6.6	3.8	9.3

(a) Draw a set of axes on graph paper. Label the horizontal axis 'Time (days)' and mark it in two-day intervals. Mark the vertical axis 'Height (cm)' and mark it in intervals of 1 cm.

(b) Carefully plot the seven points for each plant and join them up in a smooth curve. You could represent the three plants in three different colours.

(c) Write about the growth pattern of each plant.

3 This table shows how the temperature in a school kitchen varied over a 12-hour period from 05:00 to 17:00

Time	05:00	06:00	07:00	08:00	09:00	10:00	11:00
Temperature (°C)	14	14.5	16	16	18	20	21
Time	12:00	13:00	14:00	15:00	16:00	17:00	
Temperature (°C)	21.5	21	20.5	18	17	16	

(a) Draw a graph to show the change in temperature over time.

(b) What time do you think the heating came on?

(c) At what times was the temperature 19 °C?

(d) The temperature was highest between 11:00 and 13:00. Why?

(e) Why do you think the temperature dropped off quickly after 14:00?

4 This table shows the value of one euro (€1) in dollars (US$) for the first day of each month in 2013.

Month	J	F	M	A	M	J	J	A	S	O	N	D
Euro ($)	1.26	1.34	1.39	1.27	1.29	1.28	1.30	1.32	1.33	1.36	1.32	1.34

(a) Use the table to draw a graph like the one in question 8 of the last exercise.

(b) From your graph read the value of €1 in dollars on:

(i) 15 March   (ii) 15 September.

(c) From your graph, read the approximate dates when €1 was worth $1.31

⇨ Conversion graphs

Suppose you are given this currency conversion.

£1 = $1.60

You can show this on a conversion graph.

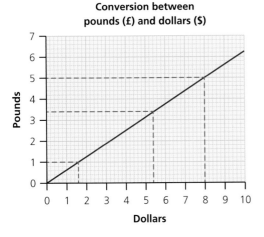

Conversion between pounds (£) and dollars ($)

Look carefully at the scale on the graph.

- Horizontally, one small square represents $0.20
- Vertically, one small square is equal to £0.20

You can read amounts off the graph:

- £3.40 is equal to $5.40
- £5 is equal to $8

Note the dotted lines that help to read off the axes accurately.

Other conversion graphs will enable you to convert quantities that are in proportion to one another.

These could be measures:

- 8 km = 5 miles

- 1 gallon = 4.5 litres

or currencies:

- £1 = $1.60

   £1 = €1.20

1 This graph shows the conversion of pounds (lb weight) to kilograms.

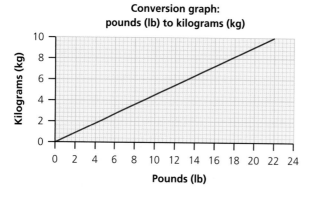

Conversion graph: pounds (lb) to kilograms (kg)

(a) What is the value of one small square on:

   (i) the horizontal axis   (ii) the vertical axis?

(b) Copy the graph and draw lines on it to convert:

   (i) 10 lb to kilograms   (ii) 4 lb to kilograms   (iii) 20 lb to kilograms.

(c) Draw lines on your graph to convert:

   (i) 10 kg to pounds (lb)   (ii) 4 kg to pounds (lb)   (iii) 6 kg to pounds (lb).

(d) Use your answer to (b)(i) to work out the value of 1 lb in kilograms.

(e) Use your answer to (c)(i) to work out the value of 1 kg in pounds (lb).

2 A good conversion is that a distance of 5 miles is equal to 8 kilometres.

   (a) What are these distances, in kilometres?

   (i) 0 miles   (ii) 10 mile   (iii) 50 miles

   (b) Draw a pair of axes on graph paper. Label the horizontal axis 'Distance in kilometres (km)' and number it from 0 to 100, using a scale of 1 cm to 10 km. Label the vertical axis 'Distance in miles' and number it from 0 to 70, using a scale of 1 cm to 10 miles.

   (c) Plot the three points from part (a) on your graph. Draw a straight line through all the points.

   (d) Write the title of your graph: 'Graph to show the conversion of kilometres to miles'.

   (e) Add straight lines on your graph to find the value of:

   (i) 70 km, in miles

   (ii) 100 km, in miles

   (iii) 20 miles, in kilometres (km)

   (iv) 45 miles, in kilometres (km).

3 You are told that £1 = 6.3 UAE dirhams (AED).

   (a) How many AED are there in:

   (i) £0   (ii) £50   (iii) £100?

   (b) Draw a pair of axes on graph paper. Label the horizontal axis 'AED' and number it from 0 to 700, using a scale of 2 cm to AED 100. Label the vertical axis 'Pounds (£)' and number it from 0 to 100, using a scale of 1 cm to £10

   (c) Plot the three points from part (a) on your graph. Draw a straight line through all the points.

   (d) Write the title of your graph: 'Graph to show the conversion of pounds sterling to AED'.

   (e) Add straight lines on your graph to find the value of:

   (i) £80, in AED

   (ii) £45, in AED

   (iii) 400 AED, in UK pounds (£)

   (iv) 75 AED, in UK pounds (£).

4 A good approximation for capacity is that 1 gallon is equal to 4.5 litres.

(a) What are these quantities, in litres?

(i) 0 gallons   (ii) 2 gallons   (iii) 10 gallons

(b) Draw a pair of axes on graph paper. Label the horizontal axis 'Litres' and number it from 0 to 50, using a scale of 2 cm to 10 litres. Label the vertical axis 'Gallons' and number it from 0 to 10, using a scale of 1 cm to 1 gallon.

(c) Plot the three points from part (a) on your graph. Draw a straight line through all the points.

(d) Write the title of your graph: 'Graph to show the conversion of litres to gallons'.

(e) Add straight lines on your graph to find the value of:

(i) 40 litres, in gallons

(ii) 25 litres, in gallons

(iii) 3 gallons, in litres

(iv) 7.5 gallons, in litres.

5 1 metre ≈ 3.3 feet

(a) What are these lengths, in feet?

(i) 0 metres   (ii) 2 metres   (iii) 10 metres

(b) Draw a pair of axes on graph paper. Label the horizontal axis 'Feet' and number it from 0 to 500. Label the vertical axis 'Metres' and number it from 0 to 150. Choose suitable scales for both axes.

(c) Draw the conversion graph and write the title.

(d) Use your graph to express:

(i) 500 ft, in metres   (iii) 75 m, in feet

(ii) 125 ft, in metres   (iv) 22 m, in feet.

20 2D shapes

⇨ Shapes

A **two-dimensional (2D) shape** is a shape that only has two dimensions (such as width or height) and no thickness. It is sometimes called a **plane figure**.

You already know about some two-dimensional shapes, such as **triangles**, **squares**, **rectangles** and other polygons.

One important thing that you need to know is the difference between a **regular** shape and an **irregular** shape.

⇨ Regular and irregular shapes

In a **regular shape** all the angles are the same and all the sides are the same length.

In an irregular shape, the angles are different and the sides are different.

Regular		Irregular
	Triangle 3 sides	
	Quadrilateral 4 sides	
	Polygon 5 or more sides	

✳ The word **polygon** is derived from the Greek words meaning 'many angles', so triangles and quadrilaterals are, strictly, polygons too.

There also many shapes that are not totally irregular. They may have some equal sides and or angles. These are often very interesting.

Here are some special triangles.

Equilateral triangle	Isosceles triangle	Right-angled triangle
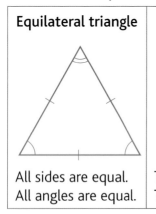	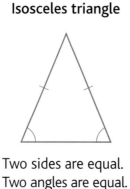	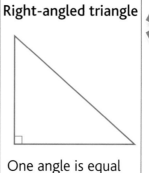
All sides are equal. All angles are equal.	Two sides are equal. Two angles are equal.	One angle is equal to 90°

*Note that equal sides are marked with small dashes and equal angles by identical curves. The right angle is a marked with a small square.

A scalene triangle has no special properties, except that the lengths of its sides are different and all its angles are different sizes.

A triangle may be obtuse-angled and isosceles, or right-angled and isosceles or right-angled and scalene.

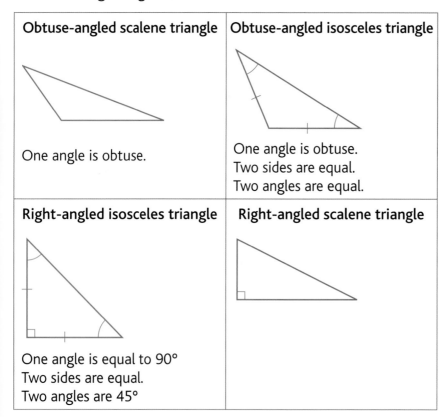

Obtuse-angled scalene triangle	Obtuse-angled isosceles triangle
One angle is obtuse.	One angle is obtuse. Two sides are equal. Two angles are equal.
Right-angled isosceles triangle	**Right-angled scalene triangle**
One angle is equal to 90° Two sides are equal. Two angles are 45°	

⇨ Parallel lines

Look at these pairs of lines.

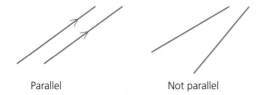

Parallel   Not parallel

Parallel lines will never meet, however long they are extended.

Railway lines are an example of parallel lines. Can you think of any others?

When you study shapes with more than three sides you will find some parallel lines. Note that they are marked with small arrows.

Here are some special quadrilaterals.

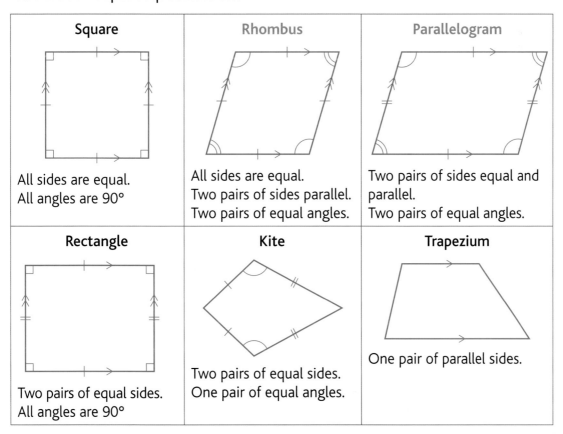

Square	Rhombus	Parallelogram
All sides are equal. All angles are 90°	All sides are equal. Two pairs of sides parallel. Two pairs of equal angles.	Two pairs of sides equal and parallel. Two pairs of equal angles.
Rectangle	**Kite**	**Trapezium**
Two pairs of equal sides. All angles are 90°	Two pairs of equal sides. One pair of equal angles.	One pair of parallel sides.

Isosceles trapezium	Isosceles arrowhead	Right-angled trapezium
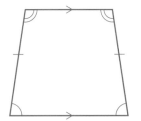	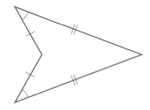	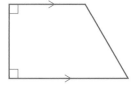
One pair of parallel sides. One pair of equal sides. Two pairs of equal angles.	Two pairs of equal sides. One pair of equal angles.	One pair of sides parallel. Two right angles.

⇨ Symmetry

A shape has a line of symmetry when you can fold it in half and the two halves match each other exactly.

The line that divides the shape in half is called a **line of symmetry**.

Exercise 20.1

Copy the triangles and quadrilaterals shown below and on the next page. Take care to include the arrows and marks that show equal and parallel sides.

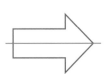
Scalene triangle

Equilateral triangle

Isosceles triangle

Right-angled scalene triangle

Obtuse-angled triangle

Obtuse-angled isosceles triangle

Right-angled isosceles triangle

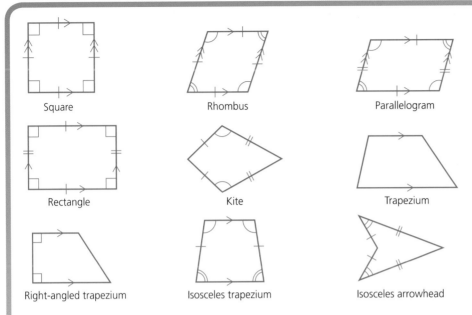

Square   Rhombus   Parallelogram

Rectangle   Kite   Trapezium

Right-angled trapezium   Isosceles trapezium   Isosceles arrowhead

1 Separate the triangles and the quadrilaterals.

2 Some of the shapes have lines of symmetry. Draw these lines on neatly with a coloured pencil. If you are not sure then fold the shape in half to check. Make sure that you have all the lines of symmetry, some have 2, one has 3 and one has 4

3 Look at the triangles. Sort them according to criteria of your choice. Discuss your ideas with a partner.

4 Now consider the quadrilaterals. Sort them according to criteria of your choice. Discuss your ideas with a partner.

5 Look again at the quadrilaterals.

   (a) Draw in their diagonals.

   (b) Under each shape, write any of these sentences that are correct.

   The diagonals are equal.

   The diagonals bisect each other.

   The diagonals meet at right angles.

   One diagonal bisects the other.

   One diagonal is a line of symmetry.

   Both diagonals are lines of symmetry.

⇨ Rotational symmetry

A shape has rotational symmetry when it can be rotated about its centre and still look the same.

Look at these shapes. Trace over them, put a pencil in the centre of each shape and then turn your tracing paper around.

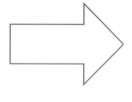  

This arrow will never look like itself until it has been rotated one full turn. It has **no** rotational symmetry.

This star will look like itself after one fifth of a turn and then again every fifth of a turn. It has rotational symmetry of order 5

This cross will look like itself after one quarter of a turn and then again every quarter of a turn. It has rotational symmetry of order 4

Exercise 20.2

1 Use tracing paper to find if these shapes have rotational symmetry. If they do, write down the order of rotational symmetry.

   (a)

   (c)

   (b)

   (d)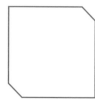

2 Copy these drawings and add shading to make patterns that have rotational symmetry of order 2. Draw in any lines of symmetry.

3 Copy these drawings and add shading to make patterns that have rotational symmetry of order 4. Draw in any lines of symmetry.

4 Copy these drawings and add shading to make patterns that have rotational symmetry of order 3. Draw in any lines of symmetry.

5 Copy these drawings and add shading to make patterns that have rotational symmetry of order 6. Draw in any lines of symmetry.

⇨ Properties of quadrilaterals

In the last exercise, you found out about the properties of quadrilaterals. The next exercise will help you to summarise what you have learnt.

Exercise 20.3

Each shape is a quadrilateral. You should now have enough information to answer the questions and identify them.

1 I have four equal sides but no right angles. What am I?

2 I have two pairs of equal angles and two pairs of equal sides. What am I?

3 I have four lines of symmetry and four sides. What am I?

4 I have four sides, one line of symmetry and two pairs of equal sides. What am I? How many pairs of equal angles do I have?

5 I have four sides, one line of symmetry and one pair of equal sides. What am I? How many pairs of equal angles do I have?

6 I have two lines of symmetry and my diagonals are equal. What am I?

7 I have no lines of symmetry but I have two pairs of parallel sides. What am I?

8 I have no lines of symmetry but I have one pair of parallel sides. What am I?

9 I have a different name but could be classified as a parallelogram. What am I?

10 I have a different name but could be classified as a rectangle or as a parallelogram. What am I?

⇨ Angles and lengths

Four of the special quadrilaterals had pairs of equal angles and no right angles.

Measure the angles of these quadrilaterals. What do you notice?

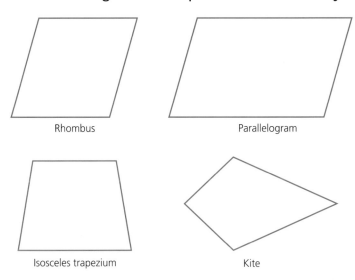

You should have noticed that some of the angles that were not equal added up to 180°. They are shown in this diagram.

Angles that are the same colour are equal.

An angle coloured pink and an angle coloured blue add up to 180°

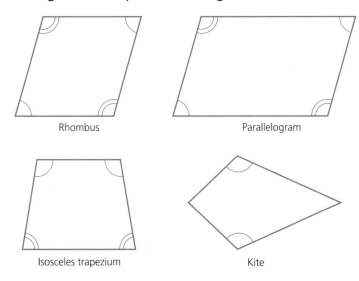

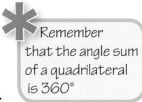

Remember that the angle sum of a quadrilateral is 360°

You can use these facts to find missing lengths and angles on shapes.

Example:

In this parallelogram, angle $A = 60°$, side $AB = 6$ cm and side $BC = 4$ cm.

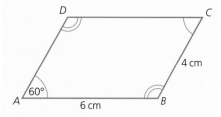

Find the size of:

(i) $\angle B$ (ii) $\angle C$ (iii) $\angle D$ (iv) side AD (iv) side CD.

(i) $\angle A + \angle B = 180°$

$60° + \angle B = 180°$

Therefore $\angle B = 120°$

(ii) $\angle C$ is opposite $\angle A$ so $\angle C = 60°$

(iii) $\angle D$ is opposite $\angle B$ so $\angle D = 120°$

(iv) AD is opposite and equal to BC so $AD = 4$ cm

(v) CD is opposite and equal to AB so $CD = 6$ cm

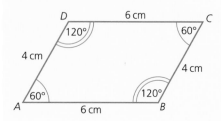

⇨ Drawing shapes accurately

In Chapter 10 you learnt how to draw triangles accurately.

Now you will find out how to draw quadrilaterals accurately.

For most of them, you will just use your ruler and protractor, as with the triangles.

You may also need to use your compasses, to mark off a length.

The kite is a bit more difficult. Look at this one.

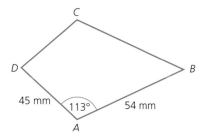

First make a sketch and mark on it all the sides and angles that you know.

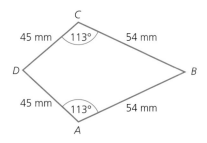

It is easier not to try to draw it as it looks here.

Start with the longest side, drawn across the page, then use a protractor to mark the angle *BAD*.

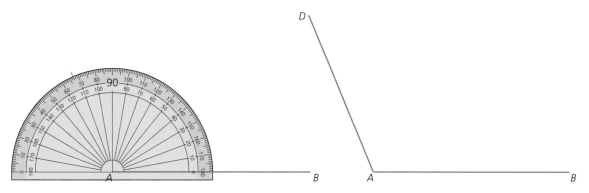

Measure the length *AD* and mark point *D*.

Use your compasses to draw two arcs to mark where *DC* and *BC* meet.

You will need to set the **radius** of the compasses to the correct length each time.

Mark point C.

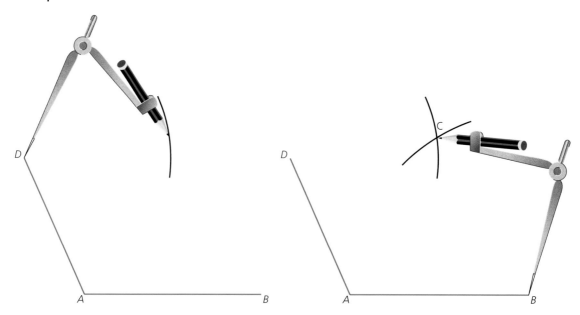

Join D to C and B to C to complete the kite.

Label the given sides and angles and measure the other angles.

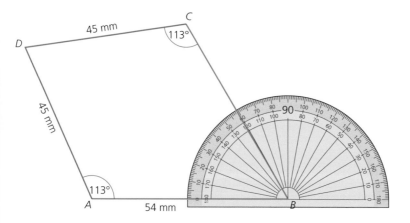

The measured angles are 75° and 59°. Check the angle sum.

113° + 113° + 75° + 59° = 360°   Correct!

20 2D shapes

Exercise 20.4

For each question, sketch the shape in your exercise book and use the properties of quadrilaterals to find the missing angles and lengths.

Label these values clearly on your shape, marking any equal or parallel lines.

Then draw the shape accurately and measure the named lengths and angles.

1 (a) In rectangle *ABCD*, write down the size of:

    (i) *AD*   **(ii)** *CD*   **(iii)** ∠*A*

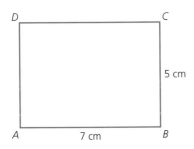

    (b) Draw *ABCD* accurately and measure the diagonals *AC* and *BD*.

2 (a) In square *PQRS*, write down the size of:

    (i) *QR*   **(ii)** *RS*   **(iii)** *PS*   **(iv)** ∠*R*

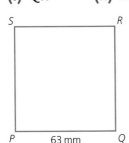

    (b) Draw *PQRS* accurately and measure the diagonals *PR* and *SQ*.

3 (a) In the isosceles trapezium *ABCD*, calculate or write down the size of:

    (i) *AD*   **(ii)** ∠*B*   **(iii)** ∠*C*   **(iv)** ∠*D*

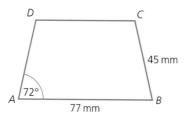

(b) Draw *ABCD* accurately and measure *CD*.

4 (a) In rhombus *PQRS*, write down the size of:

(i) *RS*   (ii) *QR*   (iii) *PS*   (iv) ∠*R*   (v) ∠*Q*   (vi) ∠*S*

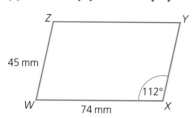

(b) Draw *PQRS* accurately and measure the diagonals *PR* and *SQ*.

5 (a) In parallelogram *WXYZ*, write down the size of:

(i) *XY*   (ii) *YZ*   (iii) ∠*W*   (iv) ∠*Y*   (v) ∠*Z*

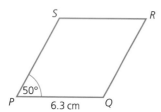

(b) Draw *WXYZ* accurately and measure the diagonals *WY* and *XZ*.

6 (a) In kite *ABCD*, write down the size of:

(i) *AB*   (ii) *AD*   (iii) ∠*A*

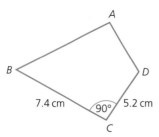

(b) Draw *ABCD* accurately and measure ∠*B* and ∠*D*.

7 (a) Sketch quadrilateral *WXYZ*.

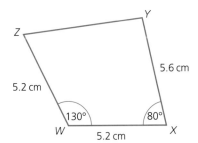

(b) Draw *WXYZ* accurately and measure:

(i) *YZ*   **(ii)** ∠*Y*   **(iii)** ∠*Z*

8 (a) In the isosceles trapezium *PQRS*, calculate and write down the size of:

(i) *PS*   **(ii)** ∠*Q*   **(iii)** ∠*P*   **(iv)** ∠*S*

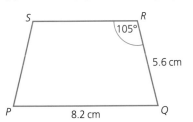

(b) Draw *PQRS* accurately and measure *RS*.

9 (a) In rhombus *WXYZ*, write down the size of:

(i) *WX*   **(ii)** *YZ*   **(iii)** *WZ*   **(iv)** ∠*W*   **(v)** ∠*Z*   **(vi)** ∠*Y*

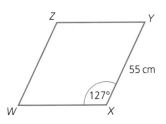

(b) Draw *WXYZ* accurately and measure the diagonals *WY* and *XZ*.

10 (a) In right-angled trapezium *ABCD*, write down the size of:

    (i) ∠*A*   **(ii)** ∠*D*   **(iii)** ∠*C*

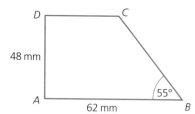

(b) Draw *ABCD* accurately and measure *CD* and *BC*.

⇨ Polygons

Polygons are 2D shapes with many sides. This means that triangles and quadrilaterals can also be called polygons.

Triangles are polygons with three sides.

Quadrilaterals are polygons with four sides.

The name comes from the Greek *polus* meaning 'many' and *gonia* meaning 'angle'. Polygons can be either regular or irregular. Regular polygons have all their sides the same length and all their interior angles equal.

The prefix *poly* also occurs in words such as:

● polyglot (a person who speaks many languages)

● polychrome (having several colours)

Regular		Irregular
	Triangle 3 sides 3 angles	
	Quadrilateral 4 sides 4 angles	

20 2D shapes

Regular		Irregular
	Pentagon 5 sides 5 angles	
	Hexagon 6 sides 6 angles	
	Heptagon 7 sides 7 angles	
	Octagon 8 sides 8 angles	
	Nonagon 9 sides 9 angles	
	Decagon 10 sides 10 angles	
	Dodecagon 12 sides 12 angles	

Polygons and symmetry

1 Trace over each of the regular polygons in the table on the previous two pages and cut them out.

2 By putting your pencil in the centre of each polygon and rotating it, work out the order of rotational symmetry of each shape.

3 By folding each shape exactly in half as many times as possible work out all the lines of symmetry of each shape.

4 Stick each shape in your exercise book. Under each one write 'This is a regular ... It has ... lines of symmetry and rotational summery of order ...'

Angles of a regular polygon

For a regular polygon, the sides are equal and the angles are equal.

For a regular triangle – an **equilateral triangle** – all the angles are 60°

For a regular quadrilateral – a **square** – all the angles are 90°

How can you find the angles of other regular quadrilaterals?

Is there a formula?

Consider the interior and exterior angles of a polygon.

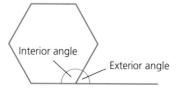

As each pair of interior and exterior angles meet on a straight line they must add up to 180°

Exercise 20.6

When you are measuring angles, you could use tracing paper to trace over the polygons and then extend the lengths of the arms of the angles so that you can measure them accurately.

1   Look at the regular polygons in the table at the beginning of this topic. Trace over the regular polygons. Extend one side of each polygon so that you have an exterior angle. Measure each exterior angle, then copy and complete the following table.

No of sides, n	Name of polygon	Size of exterior angle	$n \times$ exterior angle
3			
4			
5			
6			
7			
8			
9			
10			
12			

2   Fill in the last column of the table, the number of sides × the exterior angle.

3   Copy and complete this sentence.

The number of sides of a regular polygon multiplied by the exterior angle equals ... .

4   Rewrite the sentence like this.

The exterior angle of a regular polygon is equal to ... divided by ....

5   Add another row to the table and write a formula for the exterior angle of a regular polygon with n sides.

n			

6 Copy and complete these sentences.

The interior angle and the exterior angle of a regular polygon add up to ... .

The interior angle of a regular polygon will be equal to ... minus ... .

7 Rewrite the last sentence as a formula.

Interior angle of a regular polygon = ...

Calculating the angles of a regular polygon

You now have two formulae for the angles of a regular polygon.

Exterior angle $= \dfrac{360°}{n}$ Interior angle $= 180° - \dfrac{360°}{n}$

Now you can use these formulae to calculate angles.

Example:

Find:

(i)   the exterior angle   (ii) the interior angle

of a regular 18-sided polygon.

(i)   The exterior angle $= \dfrac{360°}{n}$

$= \dfrac{360°}{18}$

$= 20°$

(ii) The interior angle $= 180° - 20°$

$= 160°$

Even if you only need to find the interior angle, you will still have to start by finding the exterior angle.

Exercise 20.7

1 For a regular hexagon find:
   (a) the exterior angle   (b) the interior angle.

2 Find the interior angle of a regular octagon.

3 Find the interior angle of a regular decagon.

4 Find the interior angle of a regular dodecagon.

5 For a regular 18-sided polygon find:
   (a) the exterior angle   (b) the interior angle.

6 Find the interior angle of a regular 20-sided polygon.

7 Find interior angle of a regular 15-sided polygon.

8 For a regular 120-sided polygon find:
   (a) the exterior angle   (b) the interior angle.

9 For a regular 180-sided polygon find:
   (a) the exterior angle   (b) the interior angle.

⇨ Circles

If you tried to draw the last two polygons in Exercise 20.7, their exterior angles would be so small that they would probably look like this.

A polygon with a great many sides might look similar to a circle, but in fact they are different.

If you enlarged the polygon above, it would not look like a circle but a regular 16-sided shape (hexadecagon).

However much you enlarged the circle, it would still look like – and be – a circle.

Special properties of circles

Discuss with your partner what is special about a circle.

Start by thinking of a circle as a wheel.

When you have shared some ideas, check them against these properties.

- It has more lines of symmetry than you can count.
- It will always look the same, whatever angle it is rotated through.
- If you roll it, it will roll smoothly with no bumps.
- The distance from its centre to its perimeter is always the same.
- You can divide it into lots of equal slices, like a pizza.
- You can cut pieces of the sides and they will be shaped like an arch or a rainbow.
- You can draw a circle outside a regular polygon so that all the vertices or corners touch the circle.
- You can draw a circle inside a regular polygon and the midpoints of all the sides will touch the circle.

Parts of a circle

Look again at the statements above.

To put them into mathematical language, you need to be able to name the parts of a circle.

Circumference	Radius
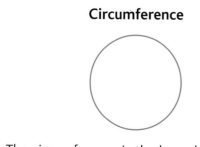	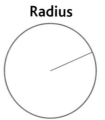
The circumference is the boundary or perimeter of the circle.	The radius is a straight line joining the centre of the circle to a point on the circumference.

Diameter	Sector

The diameter is a straight line between two points on the circumference, passing through the centre (like a diagonal).	A sector is a slice made by two radii.
Arc	**Angle at the centre**
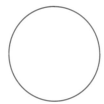	
An arc is part of the circumference of a circle.	The sum of the angles that meet at the centre is 360°

Exercise 20.8

1 If the radius of a circle is 5 cm, what is its diameter?

2 If the diameter of a circle is 24 mm, what is its radius?

3 Draw a circle with radius 4 cm and measure its diameter.

4 Draw a circle with diameter 12 cm and measure its circumference. You can use a piece of string to do this.

5 Draw four concentric circles, with these radii or diameters.

Radius 2 cm

Diameter 6 cm

Radius 5 cm

Diameter 12 cm

Concentric means having the same centre.

6 Draw a circle with diameter 10 cm and divide it into five equal sectors.

What is the angle at the centre of each sector?

7 Draw a square of side 4 cm. Now draw a circle that touches each corner of the square.

What is the diameter of the circle?

8 Copy these patterns.

Use either a pair of compasses or some graphics software on a computer.

How will you do this? Discuss with your partner how to work out where the centre of the circle will be.

(a)

The next two look simple but you may need to use some other shapes, such as squares, to find out where to put the centres of your circles.

(b)

(c)

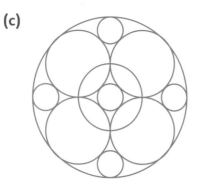

9 Now design some circle patterns of your own.

21 Pie charts and tables

⇨ Pie charts

5/5/2020

As their name suggests, **pie charts** look rather like pies that have been cut into slices.

A pie chart is a useful way of displaying information in a visual way that shows the different types of data as proportions of all the data.

Pie charts are circular and are divided into slices.

"Sector"

Understanding pie charts

My class was asked the question: 'What is your favourite subject?'

This pie chart shows the results.

You cannot tell from the pie chart how many people were asked the question. Nor can you tell how many chose each subject.

What you can do is compare the responses and identify which subjects were popular and which were not.

This pie chart shows that:

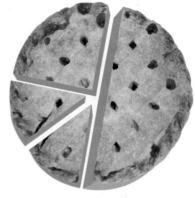

25%

¼

- Games was the most popular subject

- French was the least favourite subject

- A quarter of the class, or 25% of the pupils, liked Maths best.

■ Our favourite subjects

¹⁄₆

To find out what fraction of the class chose each subject, you need to measure the angles at the centre of the pie and do the calculations. For example, if you measure the angle at the centre of the sector for English, you will find that it is 60° ✓✓

As 60° is $\frac{1}{6}$ of 360° (a full circle) then $\frac{1}{6}$ of the class chose English as the favourite subject.

Generally, pie charts are used to compare quantities. These can be expressed as either <u>fractions</u> or <u>percentages</u>.

However, if you know how many people there are in my class, you can work out the exact numbers.

If there are 24 people in my class, then the number whose favourite subject is English is $\frac{1}{6}$ of 24

$$\frac{1}{6} \times 24 = 4$$

Exercise 21.1

1 This pie chart compares the numbers of people in the school who have packed lunch and those who have school lunch.

   (a) What percentage of the people asked have packed lunches?

   (b) What percentage have school lunch?

   (c) There are 236 pupils in the school. Work out how many of them have school lunch.

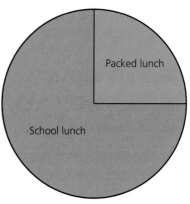

■ Lunches at my school

2 This pie chart compares the numbers of people in the school who play different musical instruments.

   (a) What fraction of the pupils play each type of instrument?

   (b) What number must you know, to be able to work out how many pupils play each type of instrument?

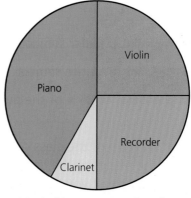

■ Musical instruments played

3 Look at this pie chart. It shows the angle at the centre of each slice.

(a) Calculate the fraction of the pie that represents each of A, B, C and D.

(b) Calculate the angle at the centre of slice E.

(c) Calculate the fraction of the pie that represents slice E.

(d) A, B, C, D and E are the number of Austrians, Bulgarians, Croatians, Danes and Estonians who took part in a survey. If a total of 120 people took part in the survey, how many were there of each nationality?

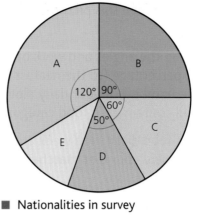
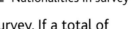
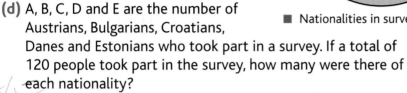

■ Nationalities in survey

4 We are going on a school trip and we have chosen our sandwich fillings for lunch. Our choices are shown on this pie chart.

(a) Measure the angles at the centre of the pie chart.

(b) Calculate the percentage of pupils who chose each sandwich filling.

(c) If 60 pupils are going on the trip, how many chose each sandwich filling?

(d) The school chef orders sandwiches in packs of five. How many packs of each type of sandwich must he order?

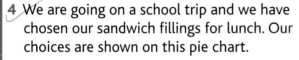
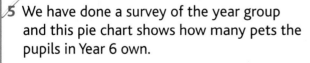

■ Sandwich fillings

5 We have done a survey of the year group and this pie chart shows how many pets the pupils in Year 6 own.

There are 36 pupils in Year 6. Use the information in the pie chart to write as many numerical facts as you can about the numbers of pets owned by pupils in Year 6

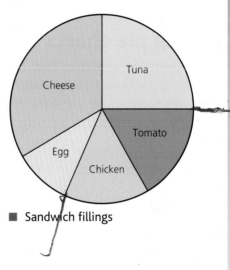
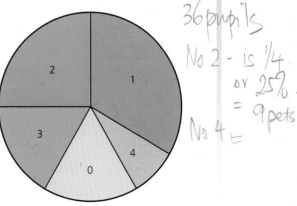

■ Numbers of pets

36 pupils
No 2 - is ¼.
or 25%.
= 9 pets
No 4 =

317

6 You may have seen this picture of an 'Eat well' plate before.

(a) Measure the angles and work out what percentage (to the nearest whole number) of your daily diet should contain each food category.

(b) Keep a food diary for a week and see how closely your diet matches the recommendations.

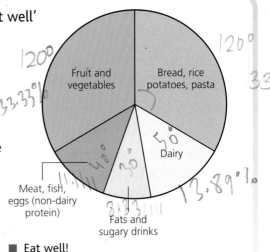

120°
33.33%
120°
33.3%
40°
30°
5°
13.89%
8x3

■ Eat well!

⇨ Drawing pie charts

You have seen from the exercises that, before you can draw a pie chart, you need to work out each category as a fraction of the whole. Then you need to use these fractions to find the angle for each sector. Then you must use a protractor to draw the chart accurately.

It helps to put your information in a table first.

Look at this table showing how the 20 pupils in my class travel to school.

How we travel to school	Tally	Frequency	Fraction of 360°	Angle
Bicycle	IIII	4	$\frac{4}{20} = \frac{1}{5}$	72°
Walk	IIII	5	$\frac{5}{20} = \frac{1}{4}$	90°
Bus	I	1	$\frac{1}{20}$	18°
Car	IIII IIII	10	$\frac{10}{20} = \frac{1}{2}$	180°
Total		20		360°

72°
$\frac{1}{5} \times 360$
$\frac{1}{4} \times 360°$

✳ The **frequency** is how often something happens or, in this case, how many people are in each category.

Draw a circle and then use your protractor to mark each angle neatly.

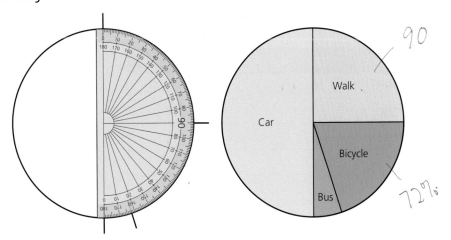

90

72%

Draw the lines to mark out the sectors.

Colour each sector and label your pie chart.

Remember to give your pie chart a title.

$\frac{1}{3}$ or $1/3$

5/5/2020

Exercise 21.2

1 Copy and complete this frequency table. Then draw a pie chart to show the information.

Favourite fruit	Tally	Frequency	Fraction of 360°	Angle
Melon	III	3	1/8	45°
Banana	HHT IIII	9	3/8	135°
Orange	IIII	4	1/6	60°
Strawberry	HHT III	8	1/3	120°
Total		24	1/1	360°

$\frac{3}{24} = \frac{1}{8} \times 360 = 45$

$\frac{9}{24} = \frac{3}{8} \times 360 = 135$

6/5/2020

2 Class 6R did a survey to see which extra subjects were the most popular. Copy and complete the frequency table and then draw a pie chart to show the information.

Extra subjects	Frequency	Fraction of 360°	Angle
Music lessons	36	1/5	72°
Sports clubs	60	1/3	120°
Ballet	45	1/4	90°
Judo	9	1/20	18°
Chess	30	1/6	60°
Total	180	1/1	360°

3 In Geography, Class 6M did a survey of transport. They counted the number of people in each car that passed the school. These are the results.

Number of people	Frequency	Fraction of 360°	Angle
1	100		
2	80		
3	24		
4	16		
More than 4	20		
Total			360°

(a) Copy and complete the frequency table and work out the angles.

(b) Draw a pie chart to show the information.

4 This is the result of a survey into the most popular tourist attractions:

Wildlife parks and zoos   5%

Historic houses   15%

Theme parks   35%

Museums and galleries   45%

Record this information in a pie chart. Follow these steps.

(a) Calculate the angle for each sector of your chart.

(b) Draw and label the pie chart.

5 My mother stocks the freezer every three months. This bar chart shows how much money she spends on each type of food.

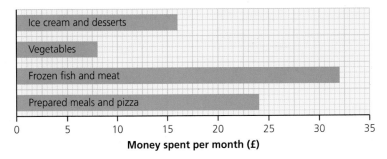

(a) Show this information in a pie chart.

(b) My father complains that the freezer is full of vegetables. Is this supported by the pie chart? How might my father's comment be explained?

⇨ Grouped data

In the exercises above the range of information fell into distinct categories. But sometimes data can take a range of values. Then you need to group the values together before you can start interpreting them.

For example, in Science, a class has been studying the effects of light, water and fertiliser on growing plants. The pupils measured, in centimetres (cm), the heights of all the plants. These are their results.

0.1	7.2	4.3	5.7	7.1	3.7	8.2	6.8	0.4	0.3
0.6	7.4	6.1	4.5	5.2	4.9	6.2	5.1	9.3	4.9
6.8	4.8	6.9	6.6	3.9	2.1	1.9	1.6	5.7	8.0

Hardly any two plants are the same height.

If you tried to draw a bar chart or pie chart, almost all the bars or sectors would be the same size. In this situation, it is sensible to group the data, like this.

Height (cm)	Tally	Total
0–0.9	IIII	4
1.0–1.9	II	2
2.0–2.9	I	1
3.0–3.9		
4.0–4.9		
5.0–5.9		
6.0–6.9		
7.0–7.9		
8.0–8.9		
9.0–9.9		
Total		30

Exercise 21.3

1 (a) Copy the grouped frequency table above and complete it.

(b) Draw a bar chart to show the information.

(c) What fraction of the plants failed to grow more than 2 cm? What do you think might have been the reason for this? Use your knowledge of Science to answer this.

(d) What fraction of the plants grew 6 cm or more? What do you think might have been the reason? Use your knowledge of Science to answer this.

2 These are the heights of the same plants a week later.

0.1	9.2	3.1	7.3	10.9	1.5	10.2	9.8	0.1	0.1
0.1	9.7	8.1	2.2	7.2	5.9	8.5	7.1	10.5	2.9
8.7	2.6	8.9	9.6	1.2	0.1	0.1	0.1	7.4	9.5

(a) Draw a grouped frequency table and complete it.

(b) Draw a bar chart to show the information.

(c) Write some facts about the results shown on the bar chart. Use scientific language to explain the results.

(d) Compare your two bar charts. Use scientific language to explain any similarities or differences.

3 These are the percentage scores of a recent Mathematics examination.

25	72	80	49	91	45
75	63	32	54	89	95
57	77	65	83	68	71

(a) Draw a grouped frequency table and complete it.

(b) Draw either a bar chart or a pie chart to show the information.

4 Here are the percentage scores of a recent Geography examination.

19	54	48	67	43	37
35	51	25	34	22	54
48	66	63	45	35	46

(a) Draw a grouped frequency table and complete it.

(b) Draw either a bar chart or a pie chart to show the information.

(c) Compare the results to the bar chart or pie chart for the Mathematics examination. Do you think the pupils have done well in their Geography examination?

(d) Write down some comments that the Geography teacher might make to the headteacher to explain the results.

Activity – Your own graphs and charts

Collect some of your own data. Use what you know about drawing charts and graphs to illustrate your findings. You may find suitable data in your Science and Geography studies too.

Here are some ideas for research.

- Most popular films, authors or TV programmes
- A climate graph for last year or the next month
- How everyone comes to school
- How many people take part in after-school activities
- How much pocket money the pupils in the class receive
- Measuring plants as they grow
- How fast ice melts

Perimeter and area

⇨ What are perimeter and area?

When you look at a plane shape, there are two dimensions you can measure.

- The distance all the way round the outside of a shape is called the **perimeter**. Imagine a sea snail, or **periwinkle**, crawling along the outside of your shape. The snail trail it leaves extends all the way around the shape. It marks out the perimeter. The perimeter is a length and is measured in millimetres, centimetres or metres. In imperial units, it is measured in inches, feet, yards or miles.

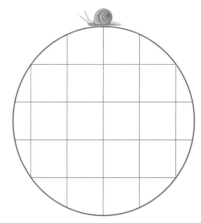

- The amount of flat space inside the shape is called the **area**. Area is measured in square units: square millimetres (mm^2), square centimetres (cm^2), square metres (m^2) or square kilometres (km^2). In imperial units, it is measured in square inches, square feet, square yards or square miles.

When discussing 2D shapes, you will also look for shapes that are **congruent** – exactly the same shape and the same size. These triangles are congruent.

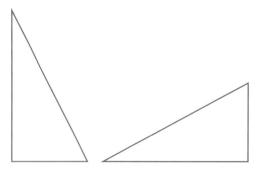

⇨ Congruent shapes

This square is divided into seven pieces that form a **tangram** puzzle.

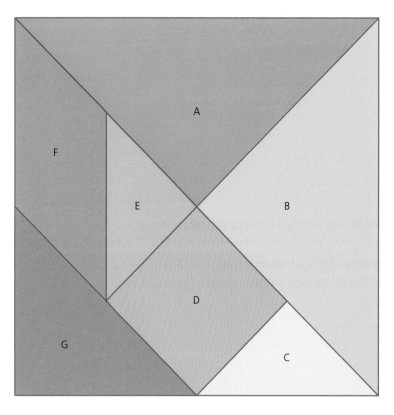

The tangram was invented in ancient China. To 'solve' the puzzle, you must arrange the pieces into a recognisable shape. The next exercise will tell you more about it.

Exercise 22.1

Copy the tangram square and cut it into seven pieces.

Use the pieces to answer these questions. Some of them may have more than one answer.

1 Which pairs of shapes are **congruent**?

2 Which shapes have the same **area**? Do any shapes that are not congruent have the same area?

3 Which shapes have the same **perimeter**? Do any shapes that are not congruent have the same perimeter?

4 Which shapes can you put together to make a square?

5 Which two shapes can you put together to make a triangle?

6 Which three shapes can you put together to make a triangle?

7 Which four shapes can you put together to make a triangle?

8 Make a parallelogram with the three smallest triangles.

9 Make a parallelogram with the five smallest pieces.

10 Make some pairs of shapes that are congruent.

11 From your shapes above write some rules about congruent shapes and their area and perimeter.

12 Make a rectangle with all the pieces.

13 Make a parallelogram with all the pieces.

14 Make a trapezium with all the pieces.

15 What can you say about the areas of your shapes in questions 12–14?

16 Make a design of your own with all the pieces.

17 Use graphics software to help you design some more exciting tangram shapes!

⇨ Perimeters and areas of squares and rectangles

You know that a **square** is a **rectangle** with **equal sides**. In the diagram, the length of the base of the rectangle is b and its height is h. The length of each side of the square is b.

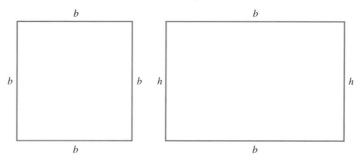

The perimeter of a square is the sum of the lengths of the four sides.

$$b + b + b + b = 4b$$

The perimeter of a rectangle is the sum of the lengths of the four sides.

$$b + h + b + h = 2b + 2h \text{ or } 2(b + h)$$

Although you can find areas by counting squares, it is easier to find the area of a rectangle by multiplying the base by the height, or the length by the width.

Written as a formula this is:

$$\text{area of a rectangle} = b \times h$$

In a square, the base is equal to the height, so the formula is:

$$\text{area of a square} = b \times b$$

or: $$\text{area of a square} = b^2$$

You can also describe the dimensions of a rectangle as length (l) and width (w), in which case the formulae would be:

$$\text{perimeter of a rectangle} = 2(l + w)$$

$$\text{area of a rectangle} = l \times w$$

You can use these formulae to find the areas and perimeters of squares, rectangles and compound shapes.

Remember the steps when using a formula.

1 Define your variables (say what the letters are and what they are worth).

2 Write down the formula.

3 Substitute numbers for the letters.

4 Complete the calculation.

5 Write down the answer, including the correct units.

When you are using formulae for 2D shapes, always sketch the shape first.

Example:

Calculate the area and perimeter of a square of side 5 cm.

Side $b = 5$ cm

Area of a square $= b^2$	Write the **formula.**
$= 5^2$	**Substitute** numbers for letters.
$= 25$ cm^2	**Calculate** then write the **answer** with the correct **units.**

5 cm

Perimeter of a square $= 4b$

$= 4 \times 5$

$= 20$ cm

In the next example the units are different.

The first thing to do is rewrite the two distances in the same units.

If the question does not specify the units to use, write the lengths in whichever unit gives the least number of 0s.

Example:

Find the perimeter and area of a rectangular field of length 1.2 km and width 80 m.

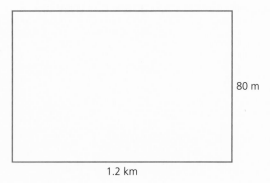

80 m

1.2 km

The dimensions of the field are given as length and width.

Length l = 1.2 km

Width w = 80 m or 0.08 km

Perimeter of a rectangle = $2l + 2w$

$$= 2 \times 1.2 + 2 \times 0.08$$

$$= 2.4 + 0.16$$

$$= 2.56 \text{ km}$$

Area of a rectangle = $l \times w$

$$= 1.2 \times 0.08$$

$$= 0.096 \text{ km}^2$$

Compound shapes

To find the perimeter of a compound shape, calculate any missing lengths and then add the lengths of all the sides.

To find the area of a compound shape, divide it into rectangles, then find the area of each rectangle and add them all together.

Example:

Find the perimeter and floor area of this L-shaped room.

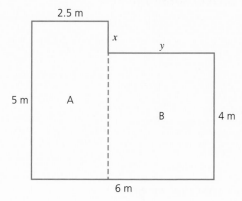

There are two missing lengths in the diagram.

They are labelled x and y.

$$x = 5 - 4 = 1 \text{ m}$$

$$y = 6 - 2.5 = 3.5 \text{ m}$$

The perimeter $= 4 + 6 + 5 + 2.5 + 1 + 3.5$

$$= 22 \text{ m}$$

The area of a rectangle is $l \times w$

For rectangle A, $l = 5$ m and $w = 2.5$ m

The area of rectangle A $= 5 \times 2.5$

$$= 12.5 \text{ m}^2$$

For rectangle B, $l = 4$ m and $w = 3.5$ m

The area of rectangle B $= 4 \times 3.5$

$$= 14 \text{ m}^2$$

The total area $=$ area of rectangle A $+$ area of rectangle B

$$= 12.5 + 14$$

$$= 26.5 \text{ m}^2$$

Exercise 22.2

1 Calculate the perimeter of each of these squares.

(a)

6 cm

(b)

50 m

(c)

1.2 km

2 Calculate the area of each of these rectangles.

(a)
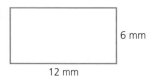
6 mm
12 mm

(b)
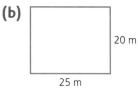
20 m
25 m

(c)
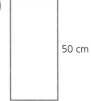
50 cm
24 cm

3 Find the area of each of these squares.

(a)

11 mm

(b)

2.5 m

(c)
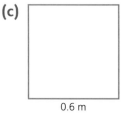
0.6 m

4 Find the perimeter of each of these rectangles.

(a)

1.2 cm
8 mm

(b)
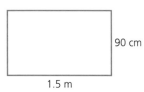
90 cm
1.5 m

(c)
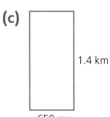
1.4 km
650 m

5 Calculate the perimeter of each of these shapes.

(a)

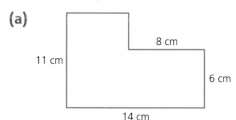

11 cm
8 cm
6 cm
14 cm

(b)

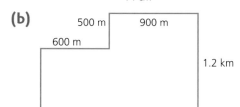

500 m
600 m
900 m
1.2 km

Remember
to work out any
missing lengths
before you start.

(c)

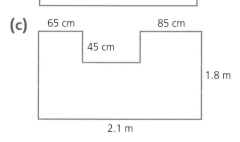

65 cm
45 cm
85 cm
1.8 m
2.1 m

6 Calculate: **(a)** the perimeter   **(b)** the area

of a rectangle measuring 3.6 m by 5 m.

7 Calculate: **(a)** the perimeter   **(b)** the area

of a square with sides 25 cm long.

8 Write down the dimensions of all the rectangles with sides that are whole numbers of metres and have area equal to 12 m².

Calculate the perimeter of each of them.

9 Write down 4 sets of dimensions for rectangles with a perimeter of 12 m.

Calculate the area of each of them.

10 This picture frame is 2 inches wide. Find the area of the frame.

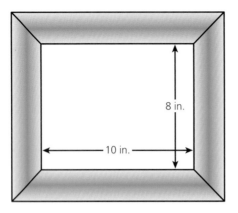

8 in.

10 in.

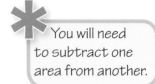

You will need to subtract one area from another.

11 A square field has sides of 750 m.

    (a) What length of fence does the farmer need, to enclose the field completely?

    (b) If fence panels are 2.5 m long, how many panels will he need?

    (c) What is the area of the field?

    (d) The farmer wants to put sheep in the field but can only have one sheep for every 4000 m^2 of land.

    How many sheep can he put in the field?

⇨ The area of a right-angled triangle

This rectangle has been divided in two congruent halves by its diagonal. Each half is a right-angled triangle.

As each right-angled triangle is half of the rectangle, then the area of one right-angled triangle must be half the area of the rectangle.

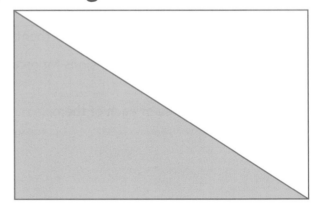

Using the formula:

area of a right-angled triangle $= \frac{1}{2} \times$ base $\times$ height (or $\frac{1}{2} \times$ length $\times$ width)

$$= \frac{1}{2} b \times h \text{ (or } \frac{1}{2} l \times w)$$

Example:

Find the area of this right-angled triangle.

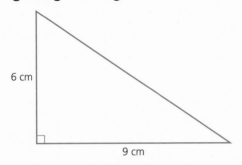

6 cm

9 cm

The area of a right-angled triangle $= \frac{1}{2} b \times h$ Write the **formula.**

$= \frac{1}{2} \times 9 \times 6$ **Substitute** numbers for letters.

$= \frac{54}{2}$ **Calculate** the answer.

$= 27 \text{ cm}^2$ Write your **answer** with the correct **units.**

Exercise 22.3

Find the area of each of these right-angled triangles.

1

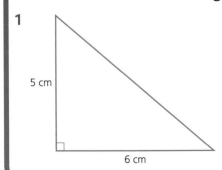

5 cm

6 cm

2

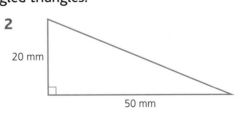

20 mm

50 mm

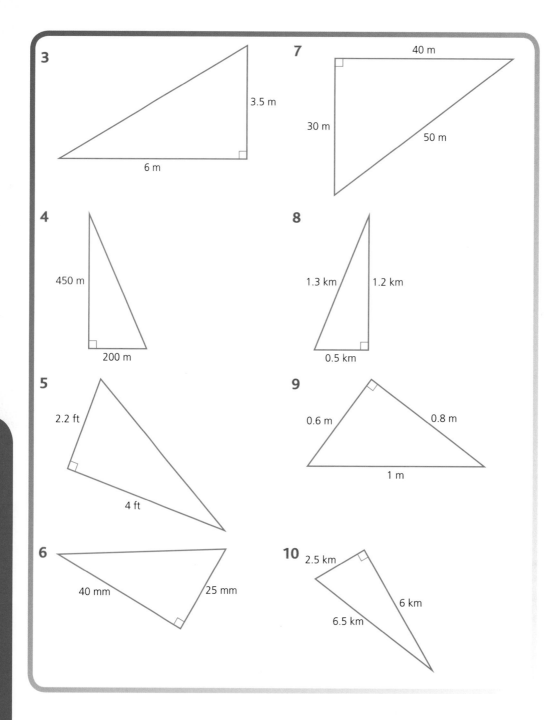

3

3.5 m

6 m

7

40 m

30 m

50 m

4

450 m

200 m

8

1.3 km

1.2 km

0.5 km

5

2.2 ft

4 ft

9

0.6 m

0.8 m

1 m

6

40 mm

25 mm

10

2.5 km

6 km

6.5 km

⇨ The area of a parallelogram

Look at this diagram.

A right-angled triangle has been cut from one end of the rectangle and moved to the other end.

What is the resulting shape?

It is a parallelogram!

Now you can write down a formula to show what has happened.

Area of parallelogram = (area of rectangle − area of triangle) + area of triangle

$$= \text{area of rectangle}$$

$$= b \times h$$

As the area of the triangle has been added and taken away, it cancels out and the net result is to leave the area of the shape unchanged.

The formula for the area of a parallelogram is the same as the formula for the area of a rectangle.

Therefore, the formula is:

The area of a parallelogram = base × height

$$= b \times h$$

Just as for a rectangle, the base and the height must be at right angles to each other.

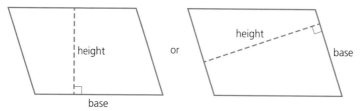

It does not matter which side is the base, as long as the height is at right angles to it!

Example:

Find the area of this parallelogram.

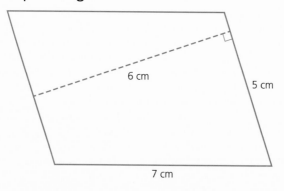

The area of the parallelogram $= b \times h$ Write the **formula.**

$\qquad\qquad\qquad\qquad\qquad = 5 \times 6$ **Substitute** numbers for the letters.

$\qquad\qquad\qquad\qquad\qquad = 30 \text{ cm}^2$ **Calculate** then write your **answer** with the correct **units.**

Exercise 22.4

1 Trace or copy each parallelogram.

    (i) Measure all the sides and record their lengths on your diagram.

    (ii) Calculate the perimeter.

    (iii) Decide which side you are going to use for the base. Draw in the corresponding height. Measure it and write down its length.

    (iv) Use the formula to work out the area of the parallelogram, giving your answer correct to the nearest square centimetre.

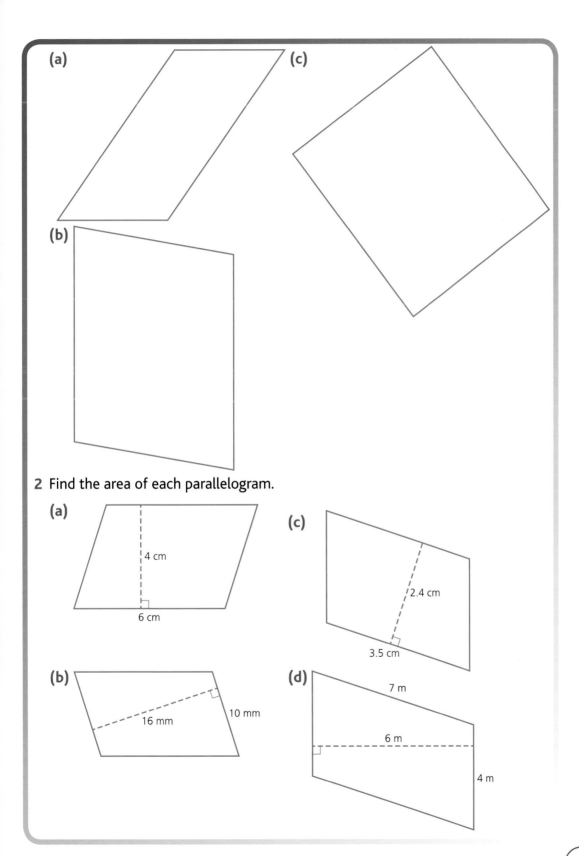

(a)

(b)

(c)

2 Find the area of each parallelogram.

(a)

4 cm

6 cm

(c)

2.4 cm

3.5 cm

(b)

16 mm

10 mm

(d)

7 m

6 m

4 m

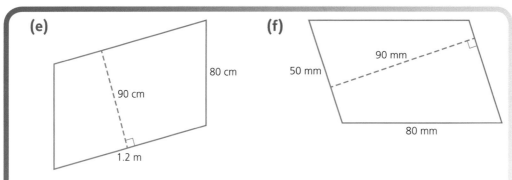

(e) 80 cm 90 cm 1.2 m

(f) 90 mm 50 mm 80 mm

3 Find the area of a parallelogram with base 5 cm and height 8 mm.

4 Find the area of a parallelogram with base 2.4 m and height 80 cm.

5 Find the area of a parallelogram with base 0.8 m and height 1.2 m.

⇨ The area of any triangle

Each of these parallelograms has been divided in two by its diagonals.

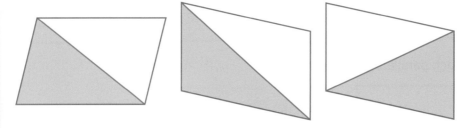

Each half of a parallelogram makes a triangle.

Therefore, based on the formula for the area of a parallelogram:

The area of a triangle $= \frac{1}{2} \times$ base $\times$ height

$$= \frac{1}{2} b \times h$$

These triangles are not right-angled, so you must find their heights before you can use the formula.

The height of any triangle is at right angles, or **perpendicular**, to its base.

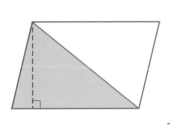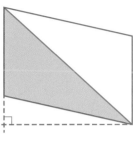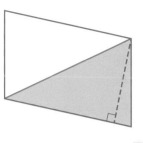

Note that to find the height of an obtuse-angled triangle, you need to extend the base until you can draw the perpendicular to the top vertex of the triangle.

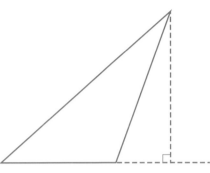

If there is not a diagram in the question you are answering, make a sketch to ensure that you have the dimensions in the correct places.

Examples:

(i) Find the area of this triangle.

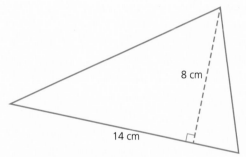

The area of a triangle $= \frac{1}{2} b \times h$ Write the **formula.**

$= \frac{1}{2} \times 14^7 \times 8$ **Substitute** numbers for the letters.

$= 56 \text{ cm}^2$ **Calculate** then write your **answer** with the correct **units.**

(ii) Find the area of triangle *ABC*, in which the base *AB* is 7 cm and the height is 6 cm.

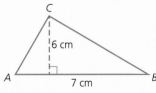

The area of $\triangle ABC = \frac{1}{2} b \times h$ Cancelling the fractions by dividing by the common factor (2) can save a stage

$= \frac{1}{2_1} \times 7 \times 6^3$ of working.

$= 21 \text{ cm}^2$

(iii) Find the area of triangle *DEF*, in which the base *DE* is 24 cm, $\angle D = 110°$ and the height *FX* is 20 cm.

The area of $\triangle DEF = \frac{1}{2} b \times h$

$$= \frac{1}{2} \times 24 \times \overset{10}{20}$$

$$= 240 \text{ cm}^2$$

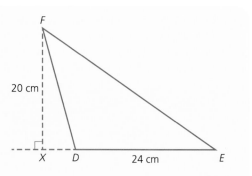

Exercise 22.5

1 Calculate the areas of these triangles

(a)

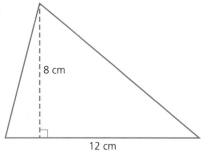

8 cm

12 cm

(b)

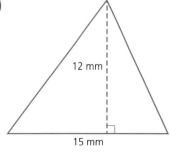

12 mm

15 mm

(c)

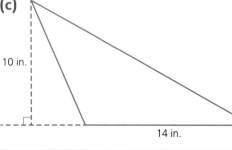

10 in.

14 in.

(d)

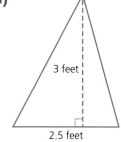

3 feet

2.5 feet

(e)

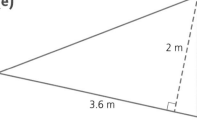

2 m

3.6 m

(f)

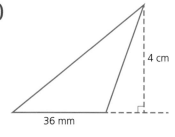

4 cm

36 mm

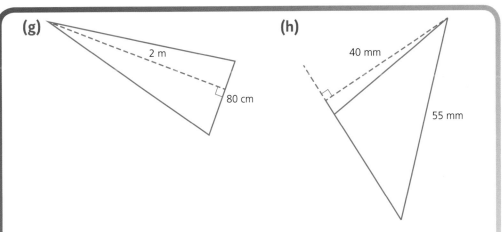

(g) 2 m, 80 cm

(h) 40 mm, 55 mm

2 Find the area of triangle *ABC* in which the base *AB* is 5 cm and the height is 8 cm.

3 Find the area of triangle *DEF* in which the base *DE* is 10 m and the height is 12 m.

4 Find the area of triangle *JKL* in which the base *JK* is 1.2 m and the height is 60 cm.

5 Find the area of triangle *DEF* in which the base *DE* is 25 mm, $\angle D = 120°$ and the height is 20 mm.

6 Find the area of triangle *XYZ* with base *XY* = 12 in, $\angle D = 115°$ and height is 10 in.

7 Find the area of triangle *PQR* with base *PQ* = 1.8 ft, $\angle D = 95°$ and height is 4 ft.

8 Find the area of triangle *ABC* with base *AB* = 3.2 m and height 2.4 m.

9 Find the area of triangle *DEF* with base *EF* = 22 mm, $\angle F = 100°$ and height is 15 mm.

10 Find the area of triangle *JKL* with base *JL* = 2.4 m, $\angle KJL = 40°$ and height *KX* = 120 cm.

The area of any triangle

343

⇨ Finding missing lengths

If you know the area of a shape, and the length of its base or height, then you can use the formula to calculate the missing length. It always helps to sketch the shape first.

Examples:

(i) A rectangle has area 24 cm^2 and height 4 cm. How long is the base?

The area of the rectangle $= b \times h$ **Formula**

$24 = b \times 4$ **Substitute**

$b = 24 \div 4$ **Calculate**

$= 6$ cm   **Answer**

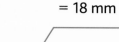

24 cm^2 4 cm

(ii) A parallelogram has area 360 mm^2 and base 20 mm. What is its height?

The area of the parallelogram $= b \times h$ **Formula**

$360 = 20 \times h$ **Substitute**

$h = 360 \div 20$ **Calculate**

$= 18$ mm   **Answer**

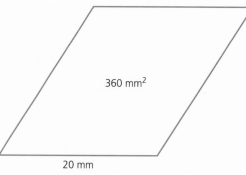

360 mm^2

20 mm

When working with a triangle formula remember that the area is half the base × height. You will therefore need to double the area before dividing.

Example:

A triangle has area 20 cm² and base 8 cm. What is its height?

The area of the triangle $= \frac{1}{2} b \times h$ **Formula**

$20 = \frac{1}{2} \times 8 \times h$ **Substitute**

$40 = 8 \times h$

$h = 40 \div 8$ **Calculate**

$= 5$ cm   **Answer**

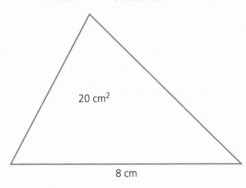

20 cm²

8 cm

Exercise 22.6

1 Find the height of each of these shapes.

(a)

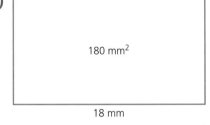

180 mm²

18 mm

(b)

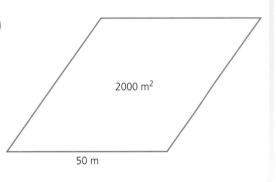

2000 m²

50 m

2 Find the length of the base in each of these shapes.

(a)

2400 m²   40 m

(b)

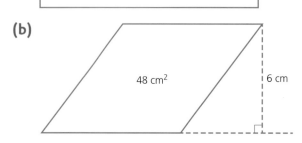

48 cm²   6 cm

3 A rectangle has area 12 cm² and height 6 cm. How long is the base?

4 A parallelogram has area 144 mm² and base 24 mm. What is its height?

5 A triangle has area 24 cm² and base 8 cm. What is its height?

6 A rectangle has area 240 m² and base 30 m. What is its height?

7 A triangle has area 2000 mm² and height 80 mm. How long is the base?

8 A parallelogram has area 132 in² and height 12 in. How long is the base?

9 A triangle has area 2.4 km² and base 1.2 km. What is its height?

10 A rectangle has area 300 cm² and height 15 cm. How long is the base?

Exercise 22.7: Summary exercise

All the shapes in this exercise are combinations of shapes of which you now know how to find the area. Divide each shape into suitable parts. Find the area of each part and add them together to find the area of the whole.

1 Find the area of this shape.

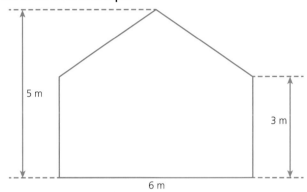

5 m

3 m

6 m

2 Find the area of this rhombus.

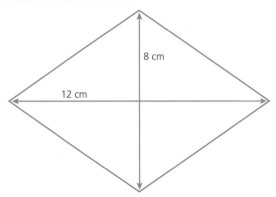

8 cm

12 cm

3 Find the area of this right-angled trapezium.

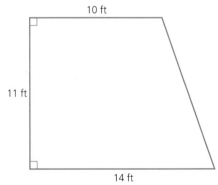

10 ft

11 ft

14 ft

4 Find the area of this kite.

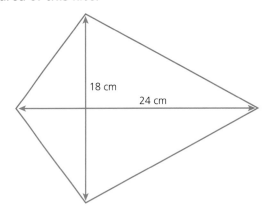

18 cm

24 cm

5 Find the area of this isosceles trapezium.

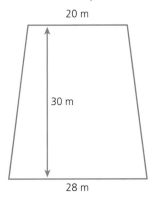

20 m

30 m

28 m

Put the two right-angled triangles from the ends together and find the area of the resulting rectangle.

6 Find the area of this isosceles arrowhead.

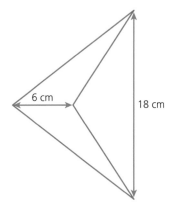

6 cm

18 cm

7 Find the area of this isosceles trapezium.

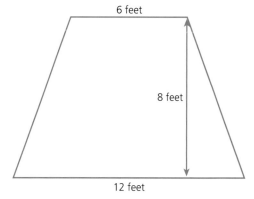

6 feet

8 feet

12 feet

8 Find the area of this regular hexagon.

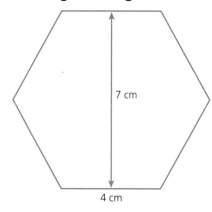

7 cm

4 cm

9 Find the area of this regular octagon.

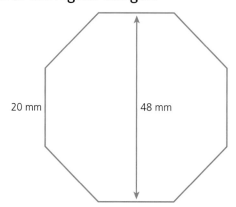

20 mm

48 mm

23 3D shapes

26/5/2020

⇨ Dimensions

The things around you are generally solid objects – they have **three dimensions**. They are called **3D** shapes. However, the shapes that you have studied so far have all been in two dimensions. It is easier to draw in two dimensions and to study angles and lengths in two dimensions.

⇨ Cuboids

Look at this three-dimensional diagram of a **cuboid**.

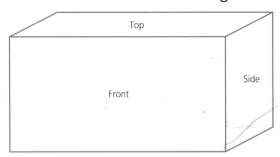

You can only see half of the cuboid, but if you picked up a solid cuboid and turned it around you would see that it has 6 faces.

This is the net of a cuboid.

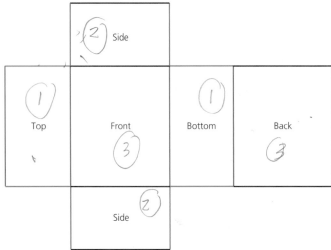

depth 3cm

width 6cm 4cm ly eight

It is similar to the net of a cube. It has six parts, one for each **face**. The faces are in identical or **congruent** pairs. The sides are the same, the top and bottom are the same and the back and front are the same.

The edges are marked in three colours: red, blue and black. When the net is folded up the outside red edges will be joined, the outside black edges will be joined and the blue edges will be joined.

26/5/2020

Exercise 23.1

1 (a) Using centimetre-squared paper, draw the net of a cuboid. Base it on the diagram on the previous page. You can choose your own dimensions but make sure that the blue lines are all the same length, the red lines are all the same length and the black lines are all the same length.

(b) Fold your cuboid up until you have a three-dimensional solid.

(c) Unfold your shape and label the faces 'top', 'bottom', 'side', 'side', 'back' and 'front'. Now fold it into a solid again.

✳ If you draw along the creases with a ball-point pen it will fold up more neatly.

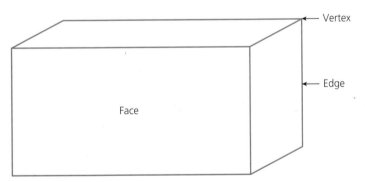

Vertex

Edge

Face

The flat sides of a solid are its **faces**, the faces meet in **edges** and the corners are its **vertices** (the plural of **vertex**).

(d) Copy and complete this sentence in your exercise book.

A cuboid has ... faces, ... edges and ... vertices.

Do you remember, from Chapter 3, what parallel lines are? A cuboid has parallel edges.

(e) Discuss with your partner where the parallel edges are on a cuboid. Use your paper cuboid to check. Are more than two edges parallel?

Copy and complete this sentence.

A cuboid has ... sets of parallel edges with ... edges in each set.

(f) We know that **perpendicular** means 'at right angles to'. Does your cuboid have perpendicular edges? Discuss your ideas with your partner.

Copy and complete this sentence.

In a cuboid, the edges that meet at a vertex are ... to each other.

2 Open out the two cuboids that you and your partner have made.

On one, colour code the pairs of parallel edges.

On the other, colour code the edges that are perpendicular to each other.

3 Close up your cuboid and fix the edges with sticky tape, to make it into a box with an opening lid.

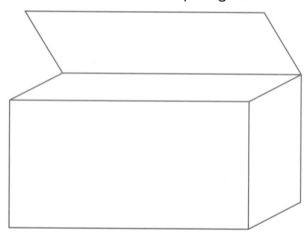

4 Fill the bottom of your cuboid with a single layer of centimetre cubes.

Copy and complete this sentence.

Inside my cuboid the bottom layer is made up of ... centimetre cubes.

Now stack centimetre cubes in one corner of your cuboid to find out how many layers you can get inside it.

Copy and complete this sentence:

My cuboid will hold ... layers of centimetre cubes.

5 Fill your cuboid with cubes, counting carefully as you go.

Copy and complete this sentence:

My cuboid contains ... centimetre cubes.

6 If you have counted correctly when you copy and complete this sentence you should have a correct statement.

The volume of my cuboid is the volume in one layer of ... $\overset{6}{...}$ centimetre cubes **multiplied by** the number of layers, ...$\overset{2}{...}$ which equals ....

$6 \times 4 \times 2 = 48 \, cm^2$

⇨ More 3D shapes   25/5

You have been looking at parallel and perpendicular edges. Faces can also be perpendicular to each other.

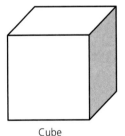

Cube

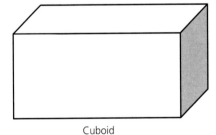

Cuboid

A cube has all the properties of a cuboid – as well as some extra properties of its own.

Here are some statements that describe a cube and a cuboid:

A cube and a cuboid both have:

● 6 faces, 8 vertices and 12 edges

● 3 pairs of parallel faces

● every face perpendicular to another face

● congruent parallel faces.

A cuboid has **three pairs of congruent rectangular faces**.

A cube has **six congruent square faces**.

⇨ Units of capacity and volume

Capacity and **volume** are both measures of how much space is taken up by something.

Capacity more usually refers to the amount of space taken up by a liquid or fluid, or the amount of space inside a hollow container.

Volume is related to the amount of space occupied by a solid.

However, in other lessons you will see that the two terms can be inter-related.

ml / liter / oz

Metric

In the metric system, you know two units of capacity.

1000 millilitres = 1 litre

When you measure objects and calculate their volume, you work in centimetres and metres, so how do these measurements relate to litres and millilitres?

This is a centimetre cube, like those you were using in the last exercise.

Its volume is 1 **cubic centimetre** (1 cm³). $1 \times 1 \times 1 = 1^3$

If you were to stack 1000 centimetre cubes into a cube 10 cm by 10 cm by 10 cm then you would have a cube of volume 1000 cm³.

$10 \times 10 \times 10 = 1000$ cm

By experiment, you will find that 1 cm³ is also equivalent to 1 millilitre.

1000 cm³ = 1 litre

Other units of volume include **cubic metres** (m³) and **cubic kilometres** (km³).

Sometimes, particularly in Europe, you will see **centilitres** used:

100 cl = 1 litre

1000 ml = 1 litre.
100 cl = 1 litre

Imperial

You will still sometimes use units from the imperial system.

2 pints = 1 quart

8 pints = 1 gallon

4 quart – 1 gallon

½ pint = ¼ quart

To convert between metric and imperial:

4.5 litres ≈ 1 gallon

1 litre = 1.75 or $1\frac{3}{4}$ imperial pints

> ✳ The American pint and the American gallon are not the same as UK imperial pints and gallons.

> ✳ It is helpful to know about the **equivalent units**, so that you can convert between metric and imperial.

Exercise 23.2

For this exercise, you will need a collection of objects of different volumes to study.

1 How big is a litre?

Work with a partner.

(a) Cut 12 straws into 10 cm lengths.

(b) Join them together with eight small balls of modelling putty to make a cube measuring 10 cm by 10 cm by 10 cm. Your cube has a volume of 1000 cm³. This is equivalent to a litre.

2 As 1 m = 100 cm

  1 m³ = 100 × 100 × 100 cm³   [handwritten above: 1m× 1m× 1]

  (a) How many cubic centimetres are there in 1 m³?   [handwritten: 1000 cm³]

  (b) How many litres are there in 1 m³?

  (c) How many cubic metres are there in 1 km³?

  (d) How many litres are there in 1 km³?

3 4.5 litres ≈ 1 gallon

  (a) Approximately how many gallons are there in 1 m³?

  (b) Approximately how many gallons are there in 1 km³?

4 1 litre = 1.75 or $1\frac{3}{4}$ imperial pints

  (a) Approximately how many pints are there in 1 m³?

  (b) Approximately how many pints are there in 5 litres?

5 Put these volumes in order, smallest first:

  10 litres   8000 ml   6 quarts   11 000 cm³

  7 pints   2 gallons   0.005 m³

6 What unit of volume would you use to measure the amount of water:

  (a) in a glass

  (b) in your bath

  (c) in a bucket

  (d) that your family uses in:

   (i) a week

   (ii) a year?

7 Now estimate each of the quantities **(a)** to **(d)** in question 6.

[handwritten notes in right margin:]
1000 cm³ = 1 litre
4.5 litres = 1 gallon
1 m³ = 1 litre – ?
100 m³ = 1 litre
10 m³ = 1 litre × 4·5
= 4·5 gallon
1 = 4 × 5 × 10
= 45

⇨ Using a formula to calculate the volume of a cuboid

You know that you can find the volume of a cuboid by finding the volume of one layer.

You also know that the volume of one layer is the area of the base of the cuboid (in cm²) multiplied by the number of centimetres in the height of the cuboid.

The volume is measured in cubic centimetres (cm³).

Since you have multiplied the base area by the number of layers, which is equal to the height, the formula for the volume of a cuboid is:

volume = length × width × height

or $V = l \times w \times h$

Remember the steps when using a formula.

1 Define your variables – say what the letters are and what they are worth.

2 Write down the formula.

3 Substitute numbers for the letters.

4 Calculate the answer.

5 Write out the answer with the correct units.

Example:

A storage container is a cuboid of length 6 m, width 2 m and height 3 m. Calculate its volume.

$l = 6$ m $\qquad w = 2$ m $\qquad h = 3$ m

Volume $= l \times w \times h$

$\qquad = 6 \times 2 \times 3$

$\qquad = 36$ m³

If the measurements in a question are given in different units, remember that you must change them all into the same units before starting any calculations.

Normally, you should use the largest unit in the question. For work on 3D shapes, it is better to use the unit that occurs most frequently.

Example:

Find the volume of a cuboid of length 4 cm, width 2 cm and height 9 mm.

$l = 4$ cm $\qquad w = 2$ cm $\qquad h = 9$ mm $= 0.9$ cm

Volume $= l \times w \times h$ $\qquad$ Write the **formula.**

$\qquad\qquad = 4 \times 2 \times 0.9$ $\quad$ **Substitute** numbers for the letters.

$\qquad\qquad = 7.2$ cm^3 $\qquad$ **Calculate**, then write your **answer** with the correct **units.**

Exercise 23.3

1 Find the volume of a cuboid with dimensions:

(a) 4 cm by 6 cm by 3 cm

(b) 5 m by 2 m by 4 m

(c) 8 mm by 10 mm by 6 mm.

2 Find the volume of a cube of side:

(a) 5 cm $\qquad$ (b) 2 m $\qquad$ (c) 12 mm.

3 Which has the greater volume, a cube with sides of 5 cm or a cuboid with sides 4 cm, 5 cm and 6 cm?

4 Find the volume of each of these cuboids.

(a) Length 2 m, width 10 cm and height 5 cm

(b) Length 5 cm, width 5 mm and height 2 m

(c) Length 3 m, width 40 cm and height 15 mm

5 A fish tank is 20 cm long, 15 cm wide and 12 cm high. What is its volume, in litres?

6 The dimensions of a terrarium are length 2.4 m, width 1.2 m and height 1.8 m. What is its volume, in litres?

7 A water tank is in the shape of a cuboid and has length 20 m, width 10 m and depth 4 m. How much water does it hold?

8 One building brick is a cube with sides of 5 cm. How many bricks will fit in a cuboid box measuring 0.4 m by 0.5 m by 1 m?

Work in centimetres.

9 A stock cube has sides of length 2 cm. How many cubes will fit inside a tin that measures 6 cm by 10 cm by 8 cm?

If your classroom is not a cuboid, find a room that is and measure that.

10 A cube of sides 80 cm is used to fill a water tank measuring 6 m by 4 m by 3 m. How many full cubes will it take to fill the tank?

11 Measure the classroom. Calculate the volume of the room.

⇨ Finding missing lengths

If you know the volume and two dimensions of a cuboid, you can find the third dimension by using **inverse** operations in the formula. The inverse of multiply is divide.

Example:

Find the height of a cuboid of length 4 cm, width 5 cm and volume 40 cm³.

Volume = $l \times w \times h$	Write the **formula.**
$40 = 4 \times 5 \times h$	**Substitute** numbers for the letters.
$40 = 20 \times h$	
$h = 40 \div 20$	**Calculate**, ÷ 20
$h = 2$ cm	Write your **answer** with the correct **units.**

10/11/2015

Exercise 23.4

1 Find the height of each of these cubes.

(a) Base area 25 cm², volume 125 cm³   25cm   5 × 5 × 5 =125

(b) Base area 36 cm², volume 216 cm³   6 × 6 × 6 = 216

(c) Length 2 cm, width 2 cm, volume 8 cm³   2 cm

(handwritten at top) $120 = 24 \times h = 5$

2 Find the height of each of these cuboids.

   (a) Base area 24 cm², volume 120 cm³ *(handwritten: 15 cm²)*

   (b) Length 3 cm, width 5 cm, volume 60 cm³ *(handwritten: 4 cm)*

   (handwritten: 25)

   (c) Length 5 cm, width 5 cm, volume 100 cm³ *(handwritten: 4 cm)*

(handwritten box:) Cuboid $V = l \times w \times h$

(handwritten:) Cube $s^3 = s \times s \times s$

3 Find the width of each of these cuboids.

   (a) Height 4 cm, length 3 cm, volume 72 cm³

   (b) Height 6 cm, length 8 cm, volume 144 cm³

   (c) Height 120 cm, length 50 cm, volume 4.8 m³   **✱** Work in metres.

4 Find the length of each of these cuboids.

   (a) Side area 15 cm², volume 60 cm³   *(handwritten:)* $V = l \times 5 \times 3 =$

   (b) Height 5 cm, width 3 cm, volume 60 cm³

   (c) Height 60 cm, width 2.5 m, volume 1.2 m³

⇨ Problem solving

You may need some interlocking cubes to make solid shapes when working through this exercise.

Exercise 23.5

1 (a) How many different cuboids can you make with 24 centimetre cubes?

   (b) How many different cuboids can you make with 36 centimetre cubes?

   (c) How many different cuboids can you make with 100 centimetre cubes?

   (d) Write down the two-digit number of cubes that will enable you to build the largest possible number of different cuboids. Briefly explain your answer.

2 (a) Build this shape from centimetre cubes. Make it hollow, so the only cubes that you use are on the faces. How many centimetre cubes are there in the shape?

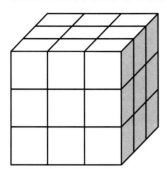

(b) Work out how many centimetre cubes you will need, to build a hollow cube with edges of:

    (i) 4 cm   **(ii)** 5 cm.

3 (a) These eight centimetre cubes are to be built into a 2 × 2 × 2 cube with red sides.

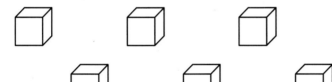

Draw the cubes and colour the sides that need to be painted red.

(b) The same eight cubes are now to be reassembled, to make a blue cube.

Copy this net of a cube and show which sides are painted **blue** and which are **red**.

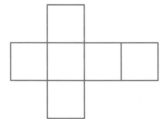

4 27 centimetre cubes are to be built into a 3 × 3 × 3 cube. Then the outside faces of the large cube are to be painted red.

(a) How many cubes with have three faces painted red?

(b) How many cubes with have two faces painted red?

(c) How many cubes with have one face painted red?

(d) How many cubes with have no faces painted red?

5 In this 3 cm by 3 cm by 3 cm cube, the cube in the middle of each face and the centre cube have been left out.

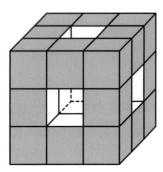

(a) How many centimetre cubes are there in the model?

(b) How many centimetre cubes would there be in a similar hollow model with edges of 4 cm?

(c) How many centimetre cubes would be in a similar hollow model with edges of 5 cm?

(d) Now work out, without building cubes, how many centimetre cubes there would be in a model with edges of:

(i) 6 cm   **(ii)** 10 cm.

(e) Write a formula for the number of centimetre cubes in a model with edges of n cm.

Activity – More nets

You have already worked on nets of cubes and tetrahedrons.

Now you are going to find out about the nets of some more 3D shapes.

Construct each net carefully, using stiff paper, ruler, pencil, protractor and compasses. When you have drawn a net, add the flaps so that you can join the edges together.

Cut out each net and score the edges with a ball-point pen.

Fold the net and glue the tabs to the appropriate faces.

1 Square based pyramid

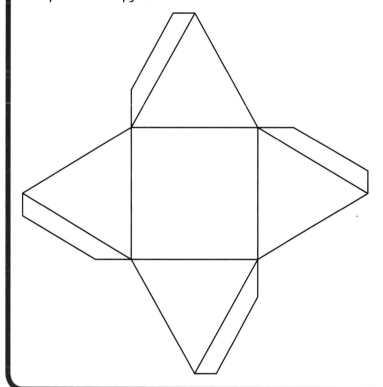

2 Hexagonal prism

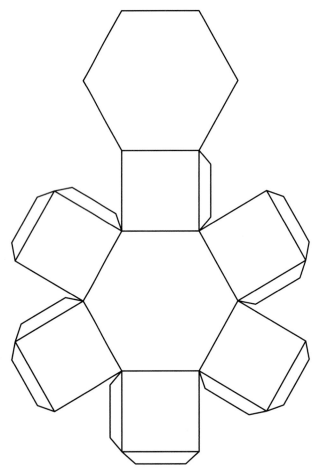

Now try some solids with curved faces.

3 Cylinder

   (a) Cut out two congruent circles of radius 3 cm.

   (b) Cut out a long rectangle of width 4 cm.

   (c) Work out how long the rectangle needs to be, so that it just wraps around the circles.

   (d) Cut the rectangle to size and stick your cylinder together.

4 Cone

(a) Cut out three congruent circles of radius 6 cm.

(b) Cut out a sector from each one, with an angle at the centre of size:

(i) 180°   (ii) 240°   (iii) 270°

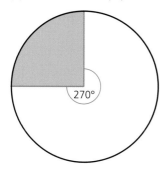

270°

(c) Take the largest sector and curl it round until the two straight edges meet. Stick your cone together. Repeat for the other two.

(d) What do you notice about the height of each cone, compared with the angle of the sector?

Averages

⇨ Representing data

In Chapters 19 and 21 you learnt about various ways of representing data in the form of graphs, charts and tables, so that you could make comparisons.

Sometimes you will need to look at information in its numerical form, rather than visually.

For example: suppose Mr Smith has marked 6B's mathematics homework.

These are their marks, out of 10.

5	8	6	4	9	7	6
7	10	7	9	7	8	7
6	9	8	8	6	7	5

Mr Smith needs to answer these questions.

- What was the highest mark?
- What was the lowest mark?
- What is the difference between the lowest and the highest mark?
- What was the average mark?
- What is the most common mark?
- What was the mark of the person who came in the middle?

In this chapter we will find out the mathematical words for some of these questions and how to calculate the answers.

⇨ The range and the mode

You can see from looking at the marks that the lowest was 4 and the highest was 10

It is less easy to see exactly what happened in between. Therefore, the next thing to do is put the information in a frequency table.

Mark	Tally	Frequency
4	I	1
5	II	2
6	IIII	4
7	IIIII I	6
8	IIII	4
9	III	3
10	I	1
Total		21

The range

The **range** is the difference between the highest value and the lowest value.

The range can be useful, because it shows how spread out the values are.

A small range would indicate that the marks were very close together.

A large range would indicate that the marks were very different.

For 6B's homework:

The range is $10 - 4 = 6$

The mode

An **average** is a value that can represent a typical value in a set of data.

The mode is the value that occurs most often in the set of data.

The mode is the value with the highest frequency. It is the only average that you can use with a set of non-numeric data.

For 6B's homework:

the mode = 7

For each set of numbers, write down:

(a) the numbers in order, smallest first

(b) the range

(c) the mode.

   1 1, 1, 2, 2, 2, 3, 3, 3, 3, 4, 4, 5

   2 6, 3, 5, 8, 7, 4, 9, 6, 5, 6, 2, 6

   3 17, 12, 13, 16, 12, 15, 16, 17, 12, 14, 16, 17

   4 20, 23, 28, 23, 25, 24, 26, 23, 24, 26, 23, 28

   5 0.5, 1.2, 0.6, 1.0, 0.9, 0.6, 1.1, 1.0, 0.9, 0.7, 0.6

   6 43, 45, 49, 48, 40, 41, 49, 42, 46, 47, 49, 44

   7 7.5, 7.8, 7.4, 7.5, 7.7, 7.8, 7.5, 7.6, 7.7, 7.5, 7.3

   8 108, 110, 112, 121, 117, 110, 120, 115, 114, 109, 111

   9 97, 72, 34, 70, 56, 91, 54, 63, 75, 83, 90

   10 54.3, 54.7, 54.7, 54.6, 54.9, 54.2, 54.8, 54.3, 54.6, 54.7

   11 A, C, B, B, C, D, A, C, B, B, E

   12 P, Q, T, S, Q, R, T, S, Q, R

⇨ The median

From your answers to questions 3 and 9 in the last exercise, you can see that the mode is not always a useful value in a set of data. Sometimes it is more useful to look for the middle value.

The median is the middle number when all the values are put in order. Looking again at 6B's homework scores, you can see that there were 21 scores. When they are placed in order, the middle score is the 11th.

List the marks, in order. Count them to find the middle.

4, 5, 5, 6, 6, 6, 6, 7, 7, 7, 7, 7, 7, 8, 8, 8, 8, 9, 9, 9, 10

Then the median score is 7

When you have an odd number of values, the median score is the one in the middle. When you have an even number of values, there are two values in the middle. To find the median you add these together and divide by 2

Examples:

Find the median of each set of numbers.

(i) 2, 5, 3, 4, 3, 6, 5, 7, 6, 4, 4

(ii) 15, 17, 16, 14, 16, 15, 16, 17, 15, 14

(i) In order, the numbers are:

2, 3, 3, 4, 4, 4, 5, 5, 6, 6, 7

There are 11 numbers.

The median is the sixth value.

The median is 4

(ii) In order, the numbers are:

14, 14, 15, 15, 15, 16, 16, 16, 17, 17

There are 10 numbers.

The median is between the fifth and sixth values.

The median is (15 + 16) ÷ 2

31 ÷ 2 = 15.5

You can also read the median from a graph.

This is the frequency bar chart for 6B's homework.

Always give your bar chart a title.

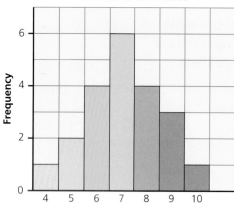

6B's homework scores

To find the median, first you need to find the total number of values.

This is 21 and thus the median is the 11th value.

Stepping along the bars and adding the frequencies:

$$1 + 2 = 3 \qquad 3 + 4 = 7 \qquad 7 + 6 = 13$$

The 11th value will be in the fourth bar.

Thus the median is 7

You can do the same with a frequency table. Look at the table for 6B's scores and add the totals. Again:

$$1 + 2 = 3 \qquad 3 + 4 = 7 \qquad 7 + 6 = 13$$

The 11th value is in the group with frequency 6

Thus the median is 7

Exercise 24.2

1 Put each set of values in order and then find the median.

(a) 5, 1, 3, 4, 1, 3, 2

(b) 10, 15, 11, 13, 12, 13, 14, 13, 12

(c) 12, 8, 99, 10, 9, 10, 11, 9, 12

(d) 23, 29, 34, 25, 45, 34, 43, 55

(e) 1.5, 0.7, 1.1, 0.8, 0.8, 1.4, 1.0, 0.8, 0.7, 1.3

(f) 45, 49, 48, 40, 41, 49, 42, 46, 47, 49

(g) 45, 54, 47, 63, 57, 48

(h) 0, 5, 9, 7, 0, 3, 2, 4

2 This frequency table shows the results of our Science investigation, when we counted the leaves on stems on a large plant.

Number of leaves on stems

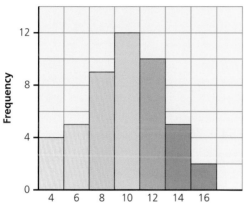

From the graph, find the median number of leaves on a stem.

3 This chart shows the amount of rainfall, in millimetres (mm), over the last month.

0	2	6	4	0	0	0	8	12	12
4	0	0	0	2	3	2	5	0	0
0	8	10	12	8	0	0	0	2	4

(a) Put the results in a frequency table.

(b) What was the median amount of rainfall?

⇨ The mean

The mean is the most commonly understood average for a set of values.

It is the **total of all the values** in a set of data, divided by the **number of values**.

$$\text{Mean} = \frac{\text{total of all the values}}{\text{number of values}}$$

Example:

These are my marks in my examinations.

Maths	81%	French	72%
English	65%	History	54%
Science	78%	Geography	76%

Calculate my mean mark.

The total of all the marks = 81 + 65 + 78 + 72 + 54 + 76

= 426

The number of marks (number of examinations) = 6

The mean mark = 426 ÷ 6

= 71%

If you have a lot of values, like Mr Smith's results for 6B's mathematics test, it is not very easy to add them all up. Instead, you can use the frequency table to help you.

Here is the frequency table again but now it has an extra column, giving the total for each mark.

It is always a good idea to check your answer.

Mark	Tally	Frequency	Total
4	I	1	4 × 1 = 4
5	II	2	5 × 2 = 10
6	IIII	4	6 × 4 = 24
7	IIII I	6	7 × 6 = 42
8	IIII	4	8 × 4 = 32
9	III	3	9 × 3 = 27
10	I	1	10 × 1 = 10
Total		21	149

24 Averages

Average mark = 149 ÷ 21

$$= 7\frac{2}{21}$$

= 7 (to the nearest whole number)

It is quite normal for the calculation of an average not to work out to one of the values in the data set, or even to a whole number. You can give a remainder as a decimal or a fraction and then round it to the nearest whole number.

Mr Smith has now worked out that for 6B's test the mean, the mode and the median were all 7

Exercise 24.3

6/5/2020

1 Calculate the mean of each set of numbers.

(a) 5, 6, 8, 7, 9

(b) 24, 25, 26, 28, 22, 25

(c) 0.2, 0.5, 1.2, 1.8, 0.8

(d) 17, 19, 18, 13, 12, 11, 16, 14

(e) 124, 106, 109, 111, 128, 145

2 What is the average overnight temperature for this week in January?

⁻5 °C, ⁻3 °C, ⁻1 °C, 1 °C, 3 °C, 2 °C, ⁻4 °C

3 What is the average of these sums of money?

24p, £3, 4p, £2, 40p, 32p

4 What is the average of these heights?

3.6 cm, 10.4 cm, 6.8 cm, 7.2 cm, 4.5 cm

5 These are the scores of a rugby team's first five matches.

10, 14, 12, 0, 24

(a) What is their mean score?

(b) Billy says that the score of 0 should not be included, because scoring nothing is not a score. Is he right?

6 A group of six pupils have measured each other's heights. These are their results.

135 cm, 142 cm, 151 cm, 165 cm, 148 cm, 138 cm

(a) What is the range of their heights?

(b) What is the mean height?

7 These are the prices of a small bottle of water in our local shops.

24p, 48p, 32p, 30p, 18p, 36p, 40p, 25p, 20p, 34p

(a) What is the range of these prices?

(b) What is the mean price? Give your answer correct to the nearest penny.

8 These figures show the number of people using the school library at lunchtime this week.

M: 25, Tu: 30, W: 24, Th: 12, F: 20

What is the mean number of people using the library each day? Give your answer correct to the nearest whole number.

9 The school cook has ordered 6 cabbages. She weighed each one and this is what she recorded.

1.1 kg, 800 g, 1.3 kg, 1.2 kg, 900 g, 850 g

What is the mean mass of her cabbages?

10 I have a five-sided spinner. This frequency table shows my scores after I spun it 30 times.

Score	Tally	Frequency	Total
1	IIII		
2	IIII IIII		
3	IIII II		
4	III		
5	IIII II		
Total			

(a) Copy and complete the table.

(b) What was my mean score?

11 This frequency table shows how many pets the pupils in my class have.

Number of pets	Tally	Frequency	Total
0	II		
1	IIII I		
2	IIII IIII		
3	II		
4	I		
Total			

(a) Copy and complete the table.

(b) What is the mean number of pets owned by pupils in the class?

12 In a science investigation we have been counting the number of seeds in apples. These are our results.

10   8   6   5   7   8   15   12   8   6

5   12   9   6   5   10   13   14   15   8

(a) Record these results in a frequency table. Add an extra column so that you can work out the totals.

(b) What is the range of the number of seeds?

(c) What is the total number of seeds?

(d) What is the mean number of seeds? Give your answer correct to the nearest whole number.

⇨ Problem solving

It is not unusual to be given the average and then have to work out the total or the number of items.

You can use the same method as for distance, speed and time. If you remember that division is the opposite of multiplication, it keeps your calculations simple.

Examples

(i) If the mean of four numbers is 6, what is the sum of the numbers?

$$\text{Mean} = \frac{\text{total of values}}{\text{number of values}}$$

$$6 = \frac{\text{sum of numbers}}{4}$$

Sum of numbers $= 4 \times 6$

$$= 24$$

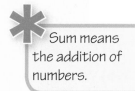

Sum means the addition of numbers.

(ii) I added together the masses of all the cakes for sale on my stall. The total was 5.6 kg. If the mean mass of a cake was 800 g, how many cakes were there?

$$\text{Mean} = \frac{\text{total of values}}{\text{number of values}}$$

$$800 = \frac{5600}{\text{number of cakes}}$$ Put both measures into grams.

$800 \times$ number of cakes $= 5600$

$$\text{Number of cakes} = \frac{5600}{800}$$ Divide top and bottom by 100

$$= 7$$

Exercise 24.4

1 The mean of five numbers is 24
   What is the sum of the numbers?

2 The class all contributed to Mr Smith's end-of-term present.
   They collected £26.40 and the mean contribution was £1.20
   How many pupils contributed to Mr Smith's present?

3 I save some money every week. I have now saved a total of £45
 If the mean amount I save is 90p, for how many weeks have I
 been saving?

4 I am growing some bean plants. I have worked out that their
 total height is 3.45 m and their mean height is 23 cm. How
 many bean plants am I growing?

5 (a) There are seven players on the netball team and, on average,
  they weigh 35 kg. What is the total mass of the team?

 (b) Mai is the reserve player. When her mass is added to the
  rest, the mean mass is 36 kg. How much does Mai weigh?

6 (a) The total mass of the class is 760 kg and the average mass
  is 40 kg. How many pupils are there in the class?

 (b) Bill was away when we did our calculation. When we
  included Bill's mass, the average dropped to 39.7 kg. What
  does Bill weigh?

7 The total of some numbers is 100. When another number is
 added, the total goes up to 120, but the mean of the numbers
 stays the same.

 (a) How many numbers were there to begin with?

 (b) What is the mean of the numbers?

8 In Science we have been counting the legs on the minibeasts we
 found in the woods. This is a frequency table of our findings. I
 worked out that the average number of legs was 6
 My science teacher asked me to check my table but I have spilt
 water on it and some numbers are missing.

Number	Tally	Frequency	Total
0	𝖳𝖧𝖫 𝖳𝖧𝖫 //		12 × 0 = 0
6	𝖳𝖧𝖫 𝖳𝖧𝖫		⎯ × 6 = 90
8	𝖳𝖧𝖫 ///		⎯ × 8 = 64
14	////		4 × 14 = 56
30	/	1	1 × 30 = 30
	Total	40	240

 (a) Copy the table and fill in the missing numbers. Show that
  my calculation is correct.

 (b) Why do you think there were more minibeasts with six legs
  than with any other number of legs?

This exercise will bring together what you have learnt.

Read through this summary before you complete the exercise.

- The **range** is the difference between the largest and smallest term.
- The **mode** is the value that occurs most often.
- The **median** is the value in the middle, when all the values are written in order.
- If there is an odd number of values, the median is the middle value.
- If there is an even number of terms, the median is the mean of the middle two values.
- The **mean** (average) is the total of all the values divided by the number of values.
- Mean $= \dfrac{\text{total of values}}{\text{number of values}}$

1 Find:

  (i) the range **(ii)** the mode **(iii)** the median **(iv)** the mean

for each set of numbers.

(a) 3, 4, 5, 5, 5, 5, 5, 6, 7

(b) 25, 23, 24, 25, 22, 21, 22, 25, 23, 25, 21, 20

(c) 0, 0.2, 0.5, 0, 0.7, 0.4, 0.2, 0, 0.8, 0, 0, 0.8

(d) 13, 85, 35, 99, 28

(e) 4.5, 1.2, 3.6, 2.4, 3.6, 2.4, 1.1, 5.8, 2.4, 1.0

2 Find the mean and range of this set of temperatures.

  $^-10\,°C$, $5\,°C$, $4\,°C$, $^-8\,°C$, $^-5\,°C$, $^-3\,°C$, $6\,°C$

3 Find the mean of these amounts of money.

  16p, £5, 80p, £12, £6, 5p, £7, 99p

4 I roll a pair of dice 40 times. These are my results.

12	4	6	9	2	11	7	5	8	10
9	10	7	9	5	8	4	7	9	5
11	6	7	3	7	10	12	6	8	7
6	7	9	8	6	8	4	3	7	6

(a) Draw a frequency table to show my results.

(b) What is the mode of these results?

(c) What is the median of these results?

(d) What is the mean value of these results? Give your answer correct to the nearest whole number.

5 This frequency table shows the results of our recent spelling test.

Number correct	Tally	Frequency	Total
14	I	1	
15	IIII	5	
16	I	1	
17	II	2	
18	III	3	
19	III	3	
20	IIII	5	
Total			

(a) Copy and complete the table.

(b) What is the range of scores?

(c) What is the modal score?

(d) What is the median score?

(e) What is the mean score?

6 This frequency table shows how many children there are in each member of the class's family.

Number of children	Frequency	Total
1	3	3 × 1 = 3
2	7	7 × 2 = 14
3	5	5 × 3 = 15
4	2	2 × 4 = 8
5	1	1 × 5 = 5
Total	18	45

(a) How many children are there in the class?

(b) What is the total number of children in all the families?

(c) What is the modal number of children?

(d) What is the median number of children?

(e) What is the mean number of children per family? Give your answer correct to one decimal place.

7 The class have raised a total of £110 for the school's nominated charity. If the mean amount donated per pupil is £4.40, how many pupils are there in the class?

8 I collected some sticks to build a model iron-age hut. I have 32 sticks and their mean length is 23 cm. What is their total length, in metres?

9 (a) There are 11 players on the football team and their mean mass is 40 kg. What is the total mass of the team?

(b) When the mass of the substitute is included, the mean mass goes down to 39.5 kg. What is the mass of the substitute?

10 The average of these numbers is 16, but two of the numbers are missing. If the range of the numbers is 7, what could the missing numbers be?

17   14   15   20   19   16

Glossary

2D   Two-dimensional; a flat shape such as a rectangle or a circle.

3D   Three-dimensional; a solid object such as a cube or a sphere.

Acute angle   An angle that is between 0° and 90°

Area   The amount of space inside a flat (2D) shape such as a rectangle or circle. It is measured in square units, such as square centimetres (cm^2).

Axes   The plural of axis. The lines at the base and to the left of a graph, which list the values of the data.

Cancel   The process of dividing the numerator (top number) and denominator (bottom number) of a fraction by one or more common factors to bring it down to its lowest terms.

Circumference   The perimeter or line round the outside of a circle.

Co-ordinates   The horizontal and vertical distances from the origin (0, 0), used to plot a point on a grid; for example, the point with co-ordinates (3, 5) is 3 units along the horizontal axis and 5 units up the vertical axis, counting from 0 on each axis.

Co-ordinate grid   The grid behind the horizontal and vertical axes.

Common factor   A number that is a factor of two or more numbers; for example, 5 is a common factor of 20 and 25

Common multiple   A number that is a multiple of two or more numbers; for example, 100 is a common multiple of 20 and 25

Composite number   A number that is not a prime number.

Conversion graph   A straight-line graph that can be used to convert between standard units of measurement or currencies.

Cube   A solid shape with six faces that are all identical squares.

Cube number   A number obtained by multiplying three 'lots' of a number together; for example, $27 = 3 \times 3 \times 3 = 3^3$

Cuboid   A solid shape with six faces that are all rectangles, although some faces may be squares.

Data   Information, for example, the highest daily temperature. Data usually comprises a group of such values that can be analysed and plotted on a chart or table.

Decimal fraction   A number less than one, written after a decimal point.

Decimal places   The place-value positions to the right of the decimal point. They are tenths, hundredths, thousandths, etc.

Denominator   The bottom number of a fraction. It tells you how many parts there are in the whole.

Diagonal A line joining non-adjacent corners on a flat shape.

Diameter A line joining two points on the circumference of a circle, that passes through the centre.

Difference The result of a smaller number being taken away from a larger number. For example, the difference between 5 and 11 is $11 - 5 = 6$

Digit A numeral, from 0, 1, 2, 3, 4, 5, 6, 7, 8, 9, used to make numbers; for example, 45 is a two-digit number.

Equilateral triangle A triangle that has three equal sides and three equal angles of 60°

Equivalent fractions Fractions that have the same value although their numerators and denominators are different; for example, $\frac{3}{4} = \frac{9}{12}$ so $\frac{3}{4}$ and $\frac{9}{12}$ are equivalent fractions.

Equivalent fractions, decimals and percentages Fractions, decimals and percentages that are equal in value; for example, $20\% = 0.2 = \frac{1}{5}$

Equivalent units The approximate comparison between metric and imperial units; for example:
1 foot ≈ 30 cm, 1 metre ≈ 3 ft 3 ins or 3.25 feet
8 kilometres ≈ 5 miles
1 lb ≈ 450 g and 1 kg ≈ 2.2 lb
1 pint ≈ 600 ml, 1 gallon ≈ 4.5 litres, 1 litre ≈ 1.7 pints
10 litres ≈ 2.2 gallons

Estimate, estimation Make an approximation, often by calculating with rounded numbers.

Factor A number that divides exactly into another number.

Factor pair A pair of factors that, when multiplied together, give the number of which they are both factors. For example, 1, 2, 3 and 6 are factors of 6 and $6 = 2 \times 3 = 1 \times 6$, so 1 and 6, 2 and 3 are factor pairs of 6

Formula A rule used to calculate a specific value, often written in letters or words; for example, the formula for the volume of a cuboid is length × width × height or $V = l \times w \times h$

Fraction A number less than one, written like this: $\frac{3}{4}$

Highest common factor The largest number that is a factor of two or more numbers; for example, 6 is the highest common factor of 24 and 30

Imperial units Non-metric units in common usage in Britain and America; for example:
Mass (weight):   16 **ounces** (oz) = 1 pound (lb)
14 **pounds** (lb) = 1 **stone** (st)   1 **ton** (t) = 2240 pounds (lb)

12 **inches** (ins) = 1 foot (ft)   3 **feet** (ft) = 1 yard (yd)
1760 **yards** (yd) = 1 **mile** (m)
2 **pints** (pt) = 1 **quart** (qt)   8 pints (pt) = 1 **gallon** (gal)

Improper fraction   A fraction in which the numerator (top number) is larger than the denominator (bottom number); for example, $\frac{7}{4}$

Integer   A whole number, it may be positive or negative; for example, 4, ⁻3 and 17 are all integers.

Inverse   The opposite. For example, the inverse of addition is subtraction.

Isosceles triangle   A triangle that has two equal sides and two equal angles.

Line graph   A line that represents the relationship between two variables, such as distance and time. It may be straight or curved. A conversion graph is also a line graph.

Long division   Division by a number with two or more digits, which shows each stage of the calculation and works down the page.

Long multiplication   Multiplication by a number with two or more digits, which shows each stage of the calculation and works out multiplication by first the units, then the tens, and so on, in each line of a frame, before the rows are added together.

Lowest common multiple   The smallest number that is a multiple of two or more numbers; for example, 50 is the lowest common multiple of 10 and 25

Lowest terms   A fraction in which the numerator and denominator have no common factors.

Mean   A typical value for a set of numerical data, equal to the sum of all the values divided by the number of values, often referred to as the 'average'; for example, the mean of 3, 5, 7 and 9 is $\frac{3+5+7+9}{4} = \frac{24}{4} = 6$

Median   The middle value in a row of numbers arranged in numerical order. If the number of values is even, the median is the mean of the middle two numbers.

Metric units   Units of mass (weight), length and volume that are in use in Britain and in Europe, as well as many other countries; for example:
1000 **milligrams** (mg) = 1 gram (g)
1000 **grams** (g) = 1 kilogram (kg)
1000 **kilograms** (kg) = 1 **metric tonne** (t)
10 **millimetres** (mm) = 1 centimetre (cm)
100 **centimetres** (cm) = 1 metre (m)

1000 millimetres (mm) = 1 metre (m)

1000 **metres** (m) = 1 kilometre (km)

1000 **millilitres** (ml) = 1 **litre** (l)

Mixed number   A combination of a whole number and a fraction; for example, $2\frac{3}{4}$

Mode   The value in a set of data that occurs most often.

Multiple   A number that is a product (multiplication) of a factor; for example, 6 is a multiple of 2, because $6 = 2 \times 3$

Negative number   A number that is less than zero, such as $^{-}4$

Numerator   The top number on a fraction. It tells you how many parts of the whole you are working with.

Obtuse angle   An angle between 90° and 180°

Obtuse-angled triangle   A triangle in which one angle is an obtuse angle.

Order of operations   The order in which a calculation should be done: brackets, index numbers, then divide, multiply, add, subtract (BIDMAS).

Parallel   Lines that are the same distance apart and will never meet, however long they are.

Parallelogram   A quadrilateral that has two pairs of equal and parallel sides.

Percentage   A fraction written in hundredths of a whole, with a percentage sign; for example, $25\% = \frac{25}{100}$

Perimeter   The distance or length around the outside of a flat (2D) shape.

Pie chart   A circular chart in which quantities can be compared by the angles at the centre of the sectors.

Polygon   A flat shape with sides that are straight lines; for example,

3 sides – a triangle	4 sides – a quadrilateral
5 sides – a pentagon	6 sides – a hexagon
7 sides – a heptagon	8 sides – an octagon
9 sides – a nonagon	10 sides – a decagon
12 sides – a dodecagon	20 sides – an icosagon

Powers of 10   The numbers that are ten multiplied by itself several times, for example:

10, 100, 1000, 10 000, 100 000, 1 000 000

Prime number   A number that has exactly two factors, itself and 1. For example, 2, 3, 5 and 7 are prime numbers, but 1 is **not** a prime number because it has only one factor.

Prime factor   Prime factors are the factors of a number that are prime numbers. For example, 2 and 3 are prime factors of 6

Product   The result of a multiplication. For example, the product of 3 and 4 is 12

Protractor   A transparent circular or semi-circular scale, used to measure angles.

Quadrant   **1** A quarter of a circle.
2 Each section of a co-ordinate grid between the horizontal and vertical axes.

Quotient   The whole-number part of an answer to a division calculation. For example, in 25 ÷ 2 = 12 remainder 1, 12 is the quotient.

Radius   A line from the centre of a circle to the circumference; in any one circle, all radii are the same length.

Rectangle   A quadrilateral with four right angles and two pairs of equal sides.

Reflection   How a shape would look if it were seen in a mirror. The resulting shape is its image.

Reflex angle   An angle that is between 180° and 360°

Remainder   The 'left over' part of a division calculation. For example, 25 ÷ 2 = 12 remainder 1

Rhombus   A parallelogram with four equal sides.

Right angle   An angle that is equal to 90°

Right-angled triangle   A triangle in which one of the angles is 90°

Rotation   The turning of an object through a number of degrees, clockwise or anticlockwise, about a centre of rotation.

Round   Write a number so that it is not exact but is a close approximation. A number can be rounded to the nearest whole number, ten, hundred thousand, etc. or to one, two, three or more decimal places. For example, 239 is 240 to the nearest ten; 3.2546 is 3.3, correct to one decimal place.

Round down   Round a number to an approximation that is smaller than the number.

Round up   Round a number to an approximation that is larger than the number.

Scalene triangle   A triangle with three sides of different lengths and three angles of different sizes.

Sector   Part of a circle between two radii.

Simplify   Reduce a fraction to its lowest terms.

Speed   A compound unit relating distance and time, usually written as kilometres per hour (km/h or kpm) or miles per hour (mph).

Square number The result of multiplying a number by itself. For example, $4 \times 4 = 16$ so 16 is a square number.

Square root A factor of a number that can be multiplied by itself to give the number. For example, $16 = 4 \times 4$, so 4 is the square root of 16

Sum The answer to an addition calculation. For example, the sum of 3 and 4 is 7

Symbol A number or operation $(+, \times, \div, -)$ written as a letter or simple shape.

Translation The description of how an object is moved, first along and then up or down, to make an image.

Vertically opposite Two of the angles formed when two straight lines cross. At every such point, there are two pairs of equal vertically opposite angles.

Index